BATTERIES & UPS IN HAZARDOUS AREAS

Ian Staff

ELECTRICAL TRAINING CONSULTANT

First Edition published 2021

2QT Limited (Publishing)

Settle

North Yorkshire

BD24 9BZ

Cover design: Dale Rennard

Images supplied by author

Printed in the UK by Lightning Source UK Limited

ISBN 978-1-914083-30-3

About the Author

I am an Electrical Training Consultant and carry out Electrical Training for a company in Hull by the name of Humberside Offshore Training Association Ltd. (H.O.T.A.), Malmo Road. Before my 15 or so years at H.O.T.A. as a Trainer/Assessor I was 38 years with BP, seven of those years as their Instrument/Electrical Training Officer in charge of all Instrument and Electrical Training in their Training Department where I obtained my Training and Assessing Qualifications.

Following on from my last books 'Hazardous Areas for Technicians', 'Inspections in Hazardous Areas', 'Motors in Hazardous Areas' and 'Earthing and Bonding in Hazardous Areas' I have produced this book 'Batteries & UPS in Hazardous Areas' which is again, aimed at Technician level.

Batteries only become a problem when something goes wrong or when tests have to be carried out and it is good to understand what systems and methods are in place and what is available.

I have included most types of battery some of which you will not have heard of and will most certainly not be in your Hazardous Areas, but it is good to see what is available and the different ways in which they work.

Ian Staff SB St J. Electrical Training Consultant.

<u>Recommended Publications Regulations & Standards for Hazardous Areas:</u>

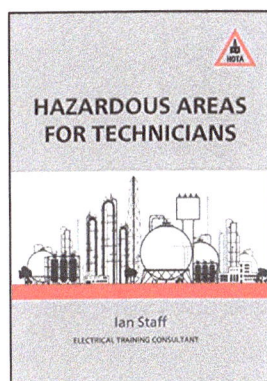

HAZARDOUS AREAS FOR TECHNICIANS	**INSPECTIONS IN HAZARDOUS AREAS**	**MOTORS & CONTROL IN HAZARDOUS AREAS**	**EARTHING & BONDING IN HAZARDOUS AREAS**
Ian Staff ELECTRICAL TRAINING CONSULTANT	Ian Staff ELECTRICAL TRAINING CONSULTANT	Ian Staff ELECTRICAL TRAINING CONSULTANT	Ian Staff ELECTRICAL TRAINING CONSULTANT
ISBN 978-1912014958	ISBN 978-1913071615	ISBN 978-1914083013	ISBN 978-1914083112

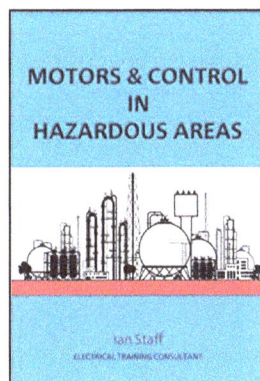

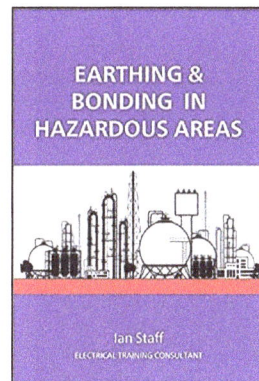

1 – BS7671: IEE Wiring Regulations   978-1785611704

2 – EEMUA: Handbook-Explosive Atmospheres   978-0859312127

3 – IEC 60079: Standard-Explosive Atmospheres   INTERNET-COST!

4 – ISO 80079: Non-Electrical/Mechanical   INTERNET-COST!

5 – DSEAR: Regulations-Explosive Atmospheres   978-0717666164

6 – 1989 Electricity at Work Regulations   978-0717666362

Contents

Battery History:

Let us have a look at how batteries were invented and how they evolved through the ages. I am sure that you will recognise many of the battery scientists and chemists as we go through history. Many batteries use Zinc and Copper Electrodes with some kind of Acid Electrolytes separated by porous pots. Many battery inventions are modifications of others. You might be amazed at the first discovery of a cell which dated back to the Parthian Empire over 2000 years ago.

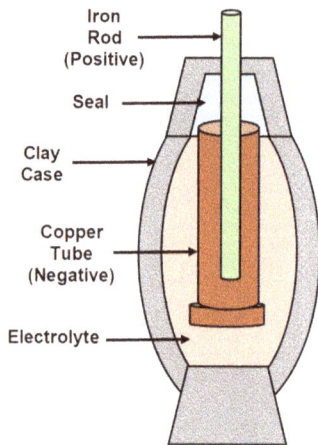

During a railway construction in Iraq in the mid 20th Century they unearthed what they thought to be one of the first if not the first battery consisting of a Copper tube, Iron rod and an electrolyte in a crude clay container. I have drawn a very simple diagram of this device on the left. It was promptly named the 'Baghdad Battery'.

In history we are going back here to the Parthian Empire who ruled in that area over 2000 years ago possibly BC. Tests were completed and it was found that when vinegar was used as the electrolyte a voltage appeared at the terminals. Now this begs the obvious question of

"What did they want a battery for?"

1744 – Ewald George von Kleist (1700 - 1748) German Physicist developed what was called the 'Leyden Jar' like the diagram to the right, but this was more like a capacitor than a battery! Now capacitors and batteries some people consider to be similar as they both give off an electric charge. Capacitors only **'STORE'** electrons with special plates whereas batteries **'MAKE'** and **'STORE'** them by converting chemical energy to electrical energy.

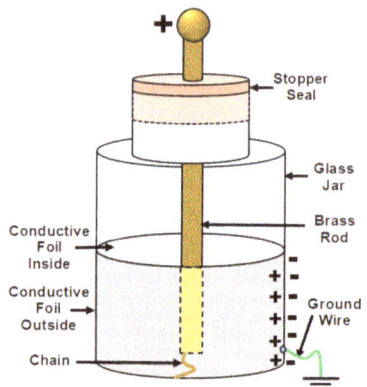

1748 – Benjamin Franklin (1706–1790) American Author and Scientist, known for his 'Kite', used the term **'Battery'** to describe his experiment using some glass plates.

1780 – Luigi Galvani (1737–1827) Italian Scientist carried out experiments on creatures such as frogs' 'twitching' limbs which he discovered by using metal links from different parts of the creature's metabolism. So he concluded that electricity was present in the creature itself. He was so close to discovering the battery but veered off on a slightly different path. His initial cell was called the Galvanic Cell in his name.

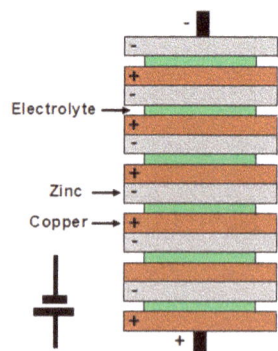

1800 – Alessandro Volta (1745 – 1827) an Italian Physicist invented the 'Voltaic Pile' as per the diagram on the right. This was claimed to be the very first **'Battery'** that could provide a constant EMF. Volta always argued with his friend Luigi Galvani (above) Volta concentrated more on dissimilar metals.

Just out of interest the battery symbol left which is used in all electrical drawings was developed from the Pile because of its shape! If you look at the Pile and then the symbol?

1790 – Eusebio Valli (1755–1816) Italian Physician made a very crude frog battery using a row of several frogs.

1800 – William Nicholson (1753–1815) English Chemist & Anthony Carlisle (1768–1840) English Surgeon discovered **'Electrolysis'.**

1801 – Johann Wilhelm Ritter (1776–1810) German Physicist invented **'Dry Storage'** batteries. He discovered the art of electroplating.

1802 – Dr William Cruickshank (Died 1810) Scottish Military Surgeon and Chemist designed the first battery to be mass produced. He laid the pates down in a trench (right) instead of putting them in a stack like Alessandro Volta's Voltaic pile! It was called the 'Trough Battery!'

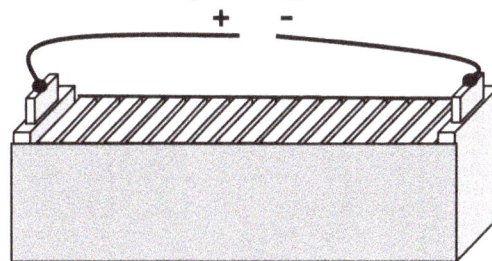

1812 – Giuseppe Zamboni (1776–1846) Italian Priest & Physicist then developed the 'Zamboni Pile' sometimes referred to as a 'Duluc Dry Pile'. This is an electrostatic battery similar to Volta's Voltaic Pile except he used silver foil, zinc foil and paper. It seems that many inventions were based on Volta's Pile.

1813 – Humphrey Davy (1778–1829) Cornish Chemist & Inventor built the first giant battery with around 2000 plates in the basement of Britain's Royal Society. This was said to be the biggest and most powerful battery in the world! Davy is the inventor of what we know in modern times as 'Electrolysis'. He was also famous for his mining Davy Lamp.

1820 – Robert Hare (1781–1858) American Chemist developed the 'Galvanic Deflagrator' which was a powerful Voltaic type battery with large plates. Hare ended up Professor of Chemistry at the University of Pennsylvania.

1834 – Michael Faraday (1791–1867) English Scientist introduced words to be used in the battery world such as **'Electrode, Electrolyte, Anode & Cathode, Anions & Cations'** which we will discuss later in the book.

1835 – William Sturgeon (1783–1850) English Physicist and Inventor coated the Zinc electrode with Mercury. Probably known more for his work on electromagnets, the basis for the electric motor.

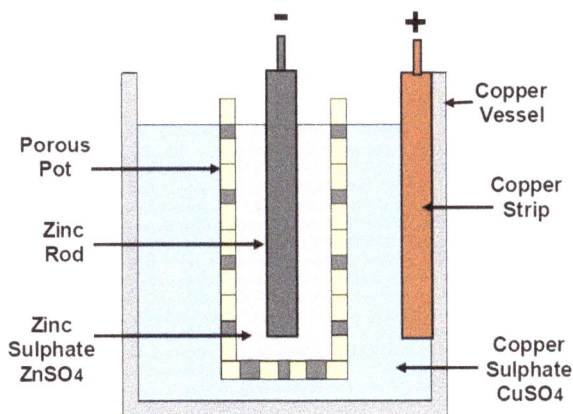

1836 – John F. Daniell (1790 – 1845) British Chemist & Meteorologist invented the 'Daniell Cell'. This included a Zinc rod immersed in Zinc Sulphate inside a porous container within the cell, which would be the negative, and Copper tape immersed in a Sulphuric Acid & Copper Sulphate Solution which would be the positive. Many descriptions show the Zinc immersed in Sulphuric Acid and not Zinc Sulphate! The Zinc electrode will gradually deplete as it oxidises and releases electrons to the Copper. Cations (explained later) will also collect on the Copper to replace the ions being used.

1837 – Golding Bird (1814–1854) A Hospital Physician developed a version of the Daniel Cell using plaster of Paris cast to separate the electrolytes.

1838 – John Dancer (1812 – 1887) English Instrument Maker developed the 'Porous Pot' Cell which was an improvement on the above Daniel Cell. It had a Zinc Anode (-) in Zinc Sulphate in a Porous Pot, (hence the name) and a Copper container filled with Copper Sulphate as the cathode (+).

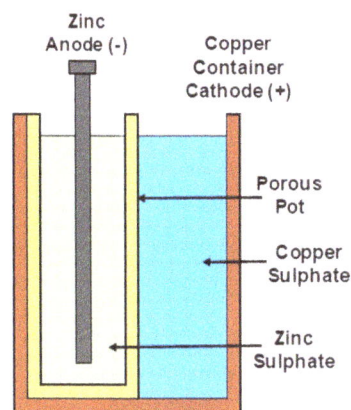

1838 – Christian Friedrich Schönbein (1799–1868) German Scientist discovered the fuel cell, but was developed later by William Robert Grove in 1842 below.

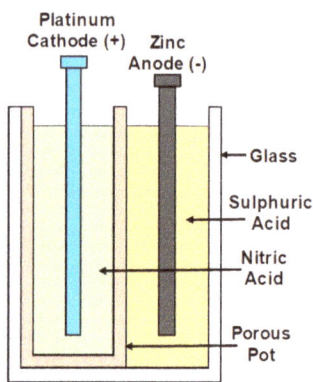

1839 – William Robert Grove (1811 – 1896) Welsh Physicist & Scientist invented the 'Grove Cell' Zinc Anode (-) in Sulphuric Acid & Platinum Cathode (+) in Nitric Acid separated by a porous container. Again a two acid electrolyte system separated by a porous pot which seems the basis for many electrolytes of this time! He also invented the first incandescent electric lamps which was enhanced later by Thomas Edison. Grove also invented a type of fuel cell which was powered by two glass tubes, one containing Oxygen and the other Hydrogen in a dilute acid electrolyte.

1839 – Edmond Becquerel (1820 - 1891) French Physicist discovered that several materials when exposed to light i.e.: Platinum plate electrodes (Pt) covered with Silver Chloride (AgCl) in a vat of acid would produce a small EMF. So he was the pioneer of course of what we call today the photoelectric effect. This was of course the basis for solar panels of today!

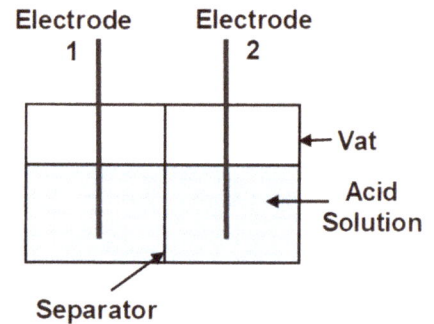

1841 – Alexander Bain (1810–1877) Scottish Inventor made a very crude Earth Battery by burying Zinc and Copper plates in the ground.

1842 – William Robert Grove (above) produced what he called his 'Fuel Cell' which he called his 'Gas Voltaic Battery'. This produced electricity by combining Hydrogen and Oxygen.

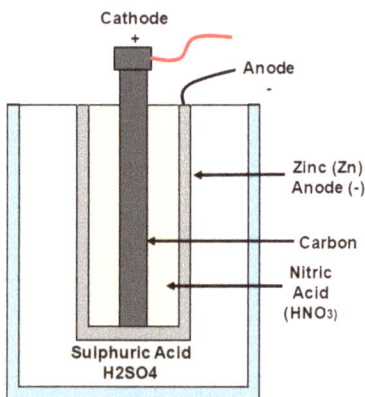

1842 – Robert Bunsen (1811 – 1899) German Chemist developed the 'Bunsen Cell' which was a Zinc anode in dilute Sulphuric Acid in a porous container and a Carbon cathode in the centre in Nitric or Chromic Acid.

1842 – Johann Christian Poggendorff (1796 – 1877) German Scientist developed the 'Poggendorff Cell'. The Cell construction was a long necked bottle as in the diagram to the right containing electrolyte and a Zinc plate between two Carbon plates.

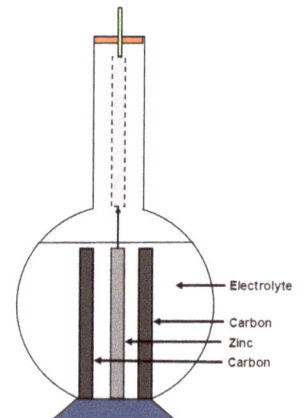

The design was slightly flawed as the electrolyte consumed the Zinc plate even if the cell was not carrying out any work so a mechanism had to be sought to lift the Zinc plate into the bottle neck when the cell was not working. He also had trouble with the single electrolyte until a guy called Eugene Grenet came along in 1959 (see later).

1845 – Carlo Matteucci (1811–1868) Italian Physicist created the frog battery. Probably the most famous, he was one of many who developed batteries using animal parts. A crude battery was made using the muscle from a frog's leg.

1853 – Isaac Lewis Pulvermacher (1815–1884) German Physicist developed the 'Pulvermacher's Chain'. Sometimes called the 'Electric Belt' was a flexible belt of linked cells that could be worn as a belt which he claimed was medical electrotherapy!

1859 – Eugene Grenet (Died 1909) Electrical Engineer and Inventor modified the single 'Poggendorff Electrolyte' (above) so the cell may be called the 'Grenet Cell'.

1859 – Heinrich Meidinger created a Power Cell called the 'Meidinger Cell'. Zinc & Copper plates in Magnesium Sulphate and Copper Sulphate.

1859 – Gaston Planté (1834–1889) French Inventor developed the first Lead Acid battery that could be recharged. This of course is the forerunner of what we use in our cars today.

1864 – Monsieur Callaud: Frenchman invented the 'Gravity Cell' which was really a variation of the Daniell Cell. Was extensively used on telegraph systems because of its increased current. Sometimes called the 'Cowfoot Cell'. Again Zinc and Copper. The Gravity name came from '**Specific Gravity**' of the electrolyte which was used in the cell. Copper electrode at the bottom, Zinc electrode at the top. Starts off with Copper Sulphate then Zinc Sulphate forms at the top. Problems keeping the electrolytes apart.

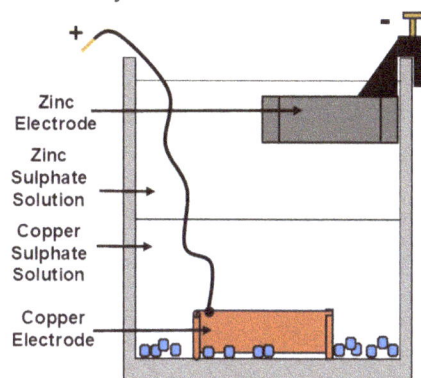

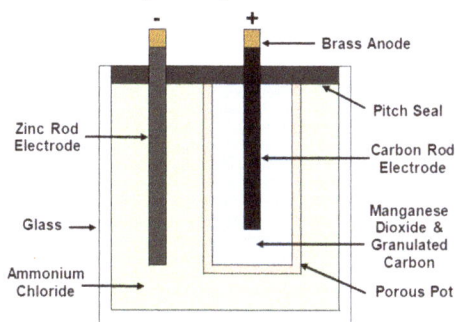

1866 – Georges Leclanché (1839 – 1882) French Physicist invented the 'Leclanché Cell', probably one of the most famous batteries that we all were taught in science. This cell is the forerunner of the dry cell we use today in torches, toys etc. The cell was a glass jar which had a Zinc rod in Ammonium Chloride electrolyte and a Carbon rod in Manganese Dioxide and Granulated Carbon electrolyte.

1869 – Pulvermacher Company developed a Pulvermacher's Chain which was a type of Voltaic battery. (Sometimes called the Electric Belt.) It was supposed to be electro-therapy.

1872 – Josia Latimer Clark (1822 – 1898) English Electrical Engineer designed a 'Clark Cell' whose output was around 1.5 volts, which stands today. We have a glass jar with Mercury in the bottom with a layer of Mercurous Sulphate above it. On top of those is a Zinc Sulphate electrolyte with a Carbon electrode, which is the cathode, running into it. At the other side of the cell is a glass rod with a Platinum wire running through it, which is the anode, dipping into the Mercury at the base. The electrodes are held up by two corks. Now Mercury batteries have been used in the 20th Century for equipment such as hearing aids & calculators, but of course Mercury is not an environmentally friendly element these days and batteries containing Mercury are no longer legal!

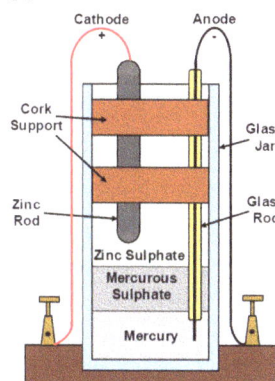

1881 – Camille Alphonse Faure (1840–1898) French Chemical Engineer invented an improved version of Planté's cell with pressed lead oxide plates and greatly increased the capacity of the battery.

1881 – J. A. Thiebaut – First Battery with both Negative Electrode & Porous Pot in a Zinc Container to make the battery lighter.

1882 – Lord Rayleigh (John William Strutt) (1842–1873) studied mathematics at Cambridge. Invented a variation of the Clark cell. This involved an 'H' Shaped vessel with Zinc Amalgam in one leg and Mercury and Mercurous Sulphate paste in the other. The electrolyte was Zinc Sulphate. In 1904 he received the Nobel Prize in physics.

1883 – Charles Fritts (1850–1903) American Inventor invented the first solar cell using Selenium and Gold.

1884 – Charles Renard (1847–1905) French Military Engineer and Arthur Constantin Krebs (1850–1935) Pioneer in Automotive Engineering, designed the very first '**Flow Battery**' which powered the La France airship.

1885 – Alfred Dunn developed a cell with Nitro-muriatic acid.

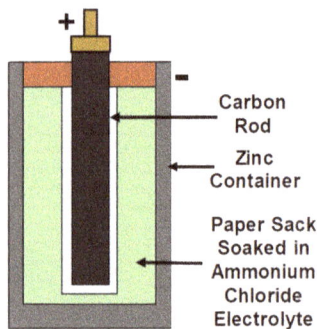

1887 – Dr Carl Gassner (1855 – 1942) German Scientist invented the first Zinc/Carbon 'Dry Cell' (Right), which together with Leclanché's Cell, is what we have today! He used Zinc container as the negative with a porous paper sack of Ammonium Chloride as the electrolyte. A Carbon rod electrode dipped in Manganese Dioxide & Carbon Black solution was used as his positive terminal. The dry cells used in torches today are only a slight modification!

1887 – Wilhelm Helleson (1836–1892) Danish inventor developed a different 'Dry Cell' with a wheat flour electrolyte which made the battery much lighter. He set up his own factory producing dry cells in 1887.

1893 – Edward Weston (1850–1936) American Chemist & Engineer developed a device with a stable voltage which was established as a 'Standard' EMF and could be used for the calibration of instruments.

1896 – Saw the first mass produced a slightly modified version of the 'Gassner Cell' by the National Carbon Company.

1899 – Waldemar Jungner (1869–1924) Swedish Inventor invented the Nickel/Cadmium 'Rechargeable Battery' which was the first alkaline battery. Also Silver/Cadmium battery.

1901 – Thomas Edison (1847–1931) American Inventor invented the 'Edison Battery' which, with Waldemar Jungner (Above), paved the way for modern alkaline batteries.

1912 – Gilbert Newton Lewis (1875–1946) American Physicist & Chemist experimented with 'Lithium Batteries'.

1932 – Dr James J. Drumm (1897–1974) Irish Chemist developed the Nickel–Zinc battery. These were actually used on railcars.

1932 – Francis Thomas Bacon (1904–1992) British Scientist. First commercial Hydrogen–Oxygen fuel cell.

1932 – Schlecht & Ackerman improved the longevity of the NiCad Battery by inventing what was to be called a 'Sintered Pole Plate' & 'Compressed Anode'.

1935 – Saw the first gel electrolyte battery.

1949 – Lewis Urry (1927–2004) Canadian Chemical Engineer & Inventor developed the small alkaline dry cell and the Lithium battery. Also Nickel–Iron batteries.

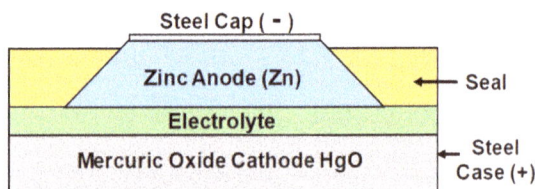

1950 – Samuel Ruben (1900 – 1988) American Inventor developed the first Zinc–Mercuric Oxide battery. Manufacturers have to be careful these days if Mercury is being used in any way. There are strict guidelines!

1954 – Gerald Pearson (1905–1987) American Physicist, Gavin Fuller (1902–1994) American Chemist & Daryl Chaplin (1906–1995) American Physicist invented the very first Solar Battery. I have a section on solar energy later in the book.

1971 – Alexandr Ilich Kloss and Boris Ioselevich Tsenter patented the first Nickel Hydrogen battery.

1980 – John Goodenough (1922–Now) Physicist invented a 'Re-chargeable Lithium Ion Battery'. He won the Nobel Prize for Chemistry in 2019. Since 1986 has been a Professor at the University of Texas.

1980s – Maria Skyllas-Kazacos (Born October 1951) Australian Chemical Engineer and Professor at the University of New South Wales pioneered the Vanadium Redox battery.

1981 – Rachid Yazami discovered 'Solid Electrolyte'. In 1985 he joined the <u>French National Centre for Scientific Research</u> as Research Associate.

1991 – Sony - Lithium Ion Battery. Lithium Ions transfer back and forth during discharge and charge. This battery is commonly used in PCs to hold the program even if the main power for the computer was to fail. There are no memory phenomena so charge and discharge can be occurring at any time interval. Lasts many years.

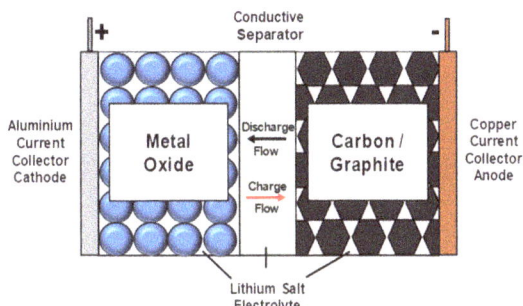

1997 – Sony – Lithium Ion Polymer battery.

Now that we have looked at batteries through history I hope that you have a better idea of the amount of inventions and famous scientists and physicists that have got us where we are today. Electrolytes and electrodes have changed enormously since the start. As we go through the book we will be looking at where the various inventions fit into today's batteries.

Battery research is into environmentally friendly materials. Glass cases and not plastic materials that are common, for instance there has been modern research into '**Urine**' batteries believe it or not and a working prototype has been produced electrolyte is free.

Experiments are going on to see what other environmentally friendly materials we can use as electrolytes. If we can make batteries from say waste fruit or vegetables what a boon that would be! Just think if manufacturers can obtain their electrolyte for free how much cheaper the batteries would be whilst helping the environment.

Let us now have a look at what materials are used to make the modern batteries we have on our factories. I am going to detail each individual component and units used in batteries and their systems i.e.: conductors, insulators, electrolytes, electrodes etc. then we will put them all together as we go through the book.

So what is a Modern Battery?

So we have gone through history and who invented what so now we ask the question "What actually is a battery?"

A battery is a device which turns energy from a **Chemical Reaction** into **DC Electrical Energy** and this would be called '**Discharge**' and in many cases we can reverse that chemical reaction back into chemical energy and this would be called '**Charge**'. Some batteries complete their discharge and are what we term as 'flat' and that is the end of their life and must be disposed of. Batteries come in many different types and in all shapes and sizes and as we continue through the book, we will discuss the different types and sizes and how the battery actually works.

Atomic Structures:

Before we can fully understand the workings of our batteries we need to know a little about the atomic structure of materials so that the formulae of the batteries can be better understood. I hope that I have not made the following explanations too complex.

Molecules:

A 'Molecule' is a small unit of an 'Element' or 'Compound'. Simply we can say that a molecule is a group of two or more 'Atoms'. The structure of a molecule is held together by what is called 'Covalent Bonding'. So for instance if we take water which has a chemical formula of H_2O, what this is telling us is that it is a molecule made up of three atoms, two of Hydrogen (H_2) and one of Oxygen (O). Carbon Dioxide CO_2 is one Carbon atom (C) bonded to two Oxygen atoms (O_2).

Another formula might include $2H_2SO_4$ (2 molecules of Sulphuric Acid) so what would be the meaning here? Well the first large 2 means that there are 2 molecules (we said that molecules contain 2 or more atoms) and the next figures and numbers show there are 7 atoms! In this case, as mentioned, the chemical symbol is for 2 molecules of Sulphuric Acid ($2H_2SO_4$) containing 2 atoms of Hydrogen (H_2), 1 atom of Sulphur (S) and 4 atoms of Oxygen (O_4).

Atoms:

So where does the above leave us with atoms? Well we can say as a definition that an atom is the smallest unit of an element or compound. Atoms are composed of a Nucleus which has Protons and Neutrons inside and Electrons in orbit around the Nucleus. The number of Electrons will always be equal to the number of Protons and this will be their **'Atomic Number'** on the **Periodic Table of Elements** with some examples below.

1	19	29	13
H	**K**	**Cu**	**Al**
Hydrogen	Potassium	Copper	Aluminium
Non Metal	Alkali Metal	Transition Metal	Post Transition

The letters on the Periodic Table left do not always seem to denote the element. It tends to pick out the older names for elements i.e.: 'K' for Potassium comes from Kalium which is Latin for Potash!

Looking at the diagram above of sections from the Periodic Table we see Hydrogen (H) has 1 Proton and 1 Electron, it is a non-metal. Potassium (K) has 19 Protons and 19 Electrons and is an Alkali Metal. Copper (Cu) has 29 Protons and 29 Electrons and is a Transition Metal, transition meaning that they are in the centre (groups 3–12) of the Periodic Table, have catalytic properties, high melting points and densities. Aluminium has 13 Protons and 13 Electrons and is Post Transition meaning it is in groups 13–15 on the Periodic Table.

Protons have a positive (+) charge, Neutrons have a neutral charge, both of these never leave the nucleus. **(Just out of interest Protons and Neutrons are made up of sub-atomic particles called Quarks!)** Electrons however, in orbit around the nucleus, will have a negative (-) charge and these, as far as our batteries are concerned, are what we are mainly interested in when discussing batteries and electron flow.

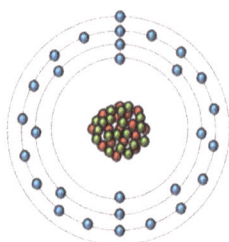

On the left is a Copper (Latin **Cu**prum Cu) Atom - Atomic Number 29. The number of Protons is always equal to the number of Electrons so our Copper Atom has 29 Protons so it will have 29 Electrons arranged in orbits and in this case there are 4 orbits. The orbits are called **'Shells'** and are arranged so the closest to the Nucleus will have up to 2 Electrons, the second orbit several Electrons and other orbits can take many more and will make up the total number of Electrons for that element. The Electrons that are the easiest to move of course and the **'loosest'** are the ones in the outer orbits or **'Valance'**. Protons and Neutrons remain locked in the Nucleus of the Atom and do not move.

What is a Conductor?

Again before we go too much further what makes a 'Conductor' different from an 'Insulator'?

Let us firstly have a look at two popular metals that are very good conductors and are used extensively in the production of electric cables and current carrying equipment.

Question:- What makes these 'Electrical Conductors' better than each other? **Answer:-** Their Atomic Structure.

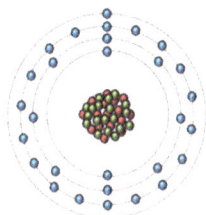

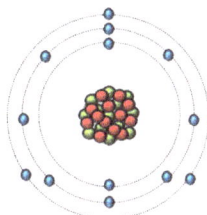

Copper
Chemical Symbol Cu
Atomic Number 29
Nucleus 29 Protons
Nucleus 34 Neutrons
4 Orbits of Electrons
29 Electrons

Aluminium
Chemical Symbol Al
Atomic Number 13
Nucleus 13 Protons
Nucleus 14 Neutrons
3 Orbits of Electrons
13 Electrons

The **Atomic Number** on the **Periodic Table** ie Copper 29 and Aluminium 13 refers to the number of Protons in the Nucleus.

The number of Protons as you will see in the diagrams on the left, is always equal to the number of Electrons. So our Copper Atom has 29 Protons and going by the above it will have 29 Electrons in 4 orbits. Aluminium has 13 Protons so will have 13 Electrons arranged in 3 orbits and so on. The orbits are called **'Shells'** and are arranged so the closest to the Nucleus will have up to 2 Electrons, the second orbit several Electrons other orbits can take many more and will make up the total number of Electrons for that element.

So if we now look at the atomic properties of these metals we find that the metallic bonds of the Electrons are very 'loose' and if a voltage is applied from, say, a battery they will freely move around the material from atom to atom, we have now got an electric current flowing. The Electrons that are the easiest to move of course and are the 'loosest' are the ones in the outer orbits or **'Valance'** which usually do not number many. The more loose Electrons in the outer or 'Valance' orbit, the easier they move. Protons and Neutrons remain locked in the Nucleus of the atom and do not move in the event of applying a voltage. When Electrons move they leave **'holes'** for others to move into.

The above metals, or conductors, are used extensively for electric cables in the electrical world. **'Copper'** is used in general cabling more than **'Aluminium'** which is not used so much in wiring equipment in domestic and industrial premises, but because it is much lighter than copper it is used for high voltage from pylon to pylon on the national grid.

Metal:	Chemical Symbol:	Atomic Number:	Protons:	Neutrons	Orbits:	Electrons:
Silver	Ag	47	47	60-62	5	47
Gold	Au	79	79	118	6	79
Zinc	Zn	30	30	35	4	30
Nickel	Ni	28	28	31	4	28
Brass	Alloy of Copper and Zinc					
Bronze	Alloy of Copper and mainly Tin					
Iron	Fe	26	26	30	4	26

The table above shows some other metals, besides Copper and Aluminium, which conduct electricity, best at the top to fair at the bottom. We do not make our cable conductors of Gold or Silver for obvious reasons although certain contacts of more expensive relays/contactors may have a Silver coating.

What is an Insulator?

We looked previously at popular conductors of electricity. Now let us look at insulators.

Question:- What makes one material an 'Insulator' and another a 'Conductor'? **Answer:-** Their atomic structure.

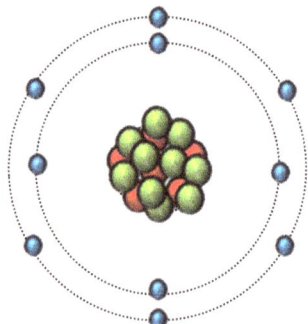

The number of Protons as you will see in the diagram on the left, is always equal to the number of Electrons. So our particular insulator atom has 10 Protons and going by the above it will have 10 Electrons and in this case we have 2 orbits.

The orbits are called **'Shells'** and are arranged so the closest to the Nucleus in our case has up to 4 Electrons and the outer orbit or 'Valance' which has 6 Electrons. Different materials will have different numbers of orbits, Electrons etc.

So if we now look at the atomic properties of these insulators we find that the bonds of the Electrons are very **'tight knit'** and if a voltage is applied from, say, a battery they will **NOT** freely move around the material from atom to atom, so there will be no electric current flowing.

In conductors, usualy metals, as mentioned, the Electrons that are the easiest to move of course and are the 'loosest' are the ones in the outer orbits or Valance which usually do not number many. Here the Electrons in the Valance do not move! Protons and Neutrons, as in the case of the conductor, remain locked in the Nucleus of the atom and do not move at all.

Several electric cable insulations are listed in the table to the right. Some you may recognise straight away. Probably PVC is the most common cable insulation as the grey twin and earth in houses would be of this insulation! Sometimes the outer sheath may be a different material to that which covers the conductors!

Initials:	Full Name:
PVC	Polyvinyl Chloride
EPL	Ethylene Propylene Rubber
XLPE	Cross Linked Polyethylene
CPE	Chlorinated Polyethylene
PTFE	Polytetrafluoroethylene

Please also remember that insulation can break down for many reasons i.e.: dampness and water. Water is **NOT** a very good conductor of electricity otherwise we could have hosepipes full of water for carrying our electricity which would be much cheaper than Copper. The fact is that water **WILL** conduct and speads all over which makes it dangerous. High voltage can also break down an insultation if high enough.

Paper	Wax
Wood	Tufnell
Nylon	Paper
Plastic	Distilled Water
Glass	Porcelain
Teflon	Oil

On the left are several more insulators, many of which you will recognise and most of these are used in the world of electricity in either equipment, batteries or cables.

As mentioned above the solid insulators on the left only do their job correctly if they are **DRY.** Get them damp and they could become conductors.

Who discovered Electrons, Protons, Neutrons:

The Electron which holds a **NEGATIVE** charge was discovered in 1897 by J. J. Thompson. The Proton has a **POSITIVE** charge and was theorised in 1815 by W. Prout. The Neutron which has **NO** charge was discovered in 1932 by J. Chadwick.

What is Electron Flow?

Electron flow occurs in two entirely different types of electricity, namely AC (**A**lternating **C**urrent) which comes out of say a 13 amp socket in your house and DC (**D**irect **C**urrent) which is of interest to us in this book at that would be our battery voltage. DC is the most efficient of the two.

So we have decided that it is Electrons that move when a voltage is put onto a conductor so it is these particles that interest us most. Electrons have a negative (-) charge, Protons have a positive (+) charge and Neutrons have no charge. Electrons move and Protons and Neutrons remain stable in the Nucleus of the atom.

Let us look at a typical Copper conductor:

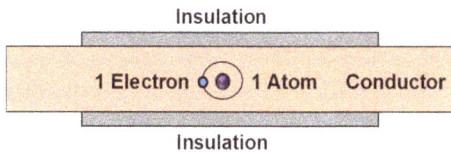

Left is a conductor which let us say is the Copper conductor of an electric cable. The grey is the insulation. I have, for ease of explanation, only drawn 1 atom with 1 Electron. Let us put a DC voltage onto the conductor and see what happens.

Our direct current (DC) voltage is going to come from a standard battery connected to either end of the conductor. You will notice the standard battery symbol below.

Looking at the diagram on the right by putting a DC voltage onto the conductor you can see that the Electron, being negatively charged, has now moved from the atom towards the positive side of the conductor. So as we mentioned the Copper conductor is made up of billions of atoms **ALL** with **'Free'** Electrons so they will **ALL** move in the same direction when the voltage is applied. So now we have electron flow, current flow is described below:

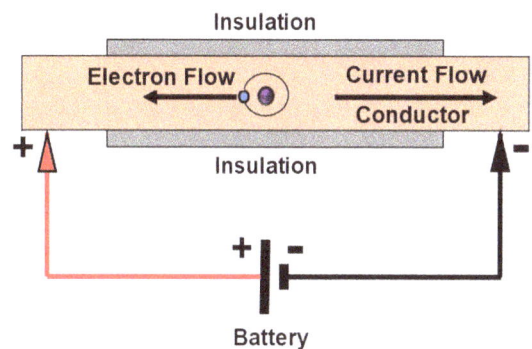

I am now going to complicate things a little! **Experiments** by **Benjamin Franklin** indicated that the current flow is in the opposite direction to the Electron flow. Now remember this is based on a theory by a great man and to tell you my opinion **'it really does not matter!'** How much current flows depends upon the voltage and what is called the 'resistance', in ohms (Ω), of our conductor.

We mentioned earlier voltage (V), amps (A) and resistance (Ω):

Amps:

Current flow is measured in amps (A) so who invented these three readings? The amp, short for **Ampere**, is named after **André-Marie Ampère** who was a French Physicist and Mathematician. If we looked at an equivalent water system then we could actually for our analogy take the water flow as the amperage.

Volts:

We have mentioned volts and they are named after **Alessandro Volta,** an Italian Physicist. So in our analogy the pressure behind the water and the force that is pushing it through the piping would be the voltage.

Ohms:

The resistance of the conductor would be measured in ohms (Ω) named after **Georg Simon Ohm** who was a German Physicist. Now here in our analogy we could say that any restrictions in the pipework i.e.: the pipe narrowing or going through a filter etc. would be our resistance.

Electrolytic Cells?

Before we go too much further and talk about electrolytes I would like to say that there are two totally different types of cell/battery that use an electrolyte. I would like to discuss these cells as they have many very different purposes and work in different ways:-

Electrolytic Cells are the units that can be used in **Electrolysis (a non-spontaneous Redox Reaction)** and require an external source of energy or in other words require a DC power supply. The **Anode** in these cells is in fact marked as the **Positive** electrode and the **Cathode** marked as the **Negative** electrode as per the diagram right. (Electrodes may be made from active materials such as Copper, Silver or Zinc or inert materials like Silicon or Graphite). This may seem a bit confusing but bear with me. Let us say that we want to Silver electroplate a piece of metal. We make the Silver bar the anode **(Positive in this type of Cell)** and the metal to be Silver plated the cathode **(Negative in this case)** and connect a battery and the process will begin.

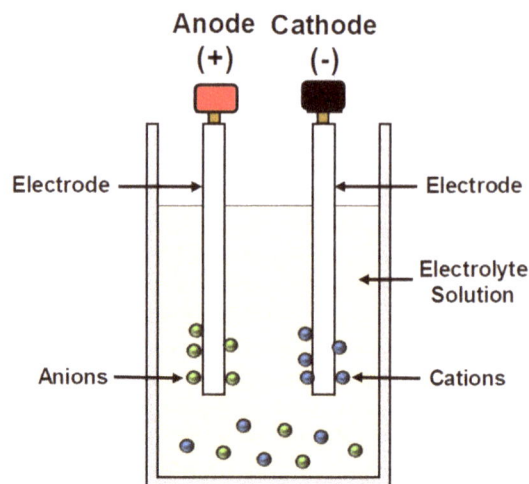

Now particles called **ions** (explanation below), some of which are called Cations which have more Protons than Electrons so they are positive (+) and others which are called Anions which have more Electrons than Protons so they are negative (-), gather on each electrode.

Cations would be formed from, for instance: Ammonia, Sodium and Iron and **Anions** would be formed from, for instance: Bromide, Sulphate and Chloride. Anions being **Negative** are attracted **HERE** to the **Positive** and Cations being **Positive** are attracted **HERE** to the **Negative**.

So we are saying that when the number of Electrons equal the number of Protons this would be an atom that is neutral. If the number of Electrons do not equal the number of Protons then the atom is not neutral, but is positively or negatively charged and is no longer an atom, IT IS AN ION! Hence we get the term **Ionisation**.

Uses: electrolytic cells can be used for:

1) Extracting gas from water: H_2O is the chemical symbol, so two gases are involved, Hydrogen & Oxygen and these can be extracted using electrolysis.
2) Electroplating: as we have already mentioned metals can be coated using electrolysis.
3) Production of high purity metals.
4) Electro-winning/Electro-refining: Extracting metal from ore i.e.: Aluminium extraction from Bauxite.

To sum up: Positively charged (having fewer Electrons) would be a **Cation** and negatively charged (having less Protons) would be an **Anion**.

So now that we have looked at the difference we will leave electrolytic cells behind and concentrate on what the book is about which are **Galvanic Cells**. Galvanic Cells, sometimes called **'Voltaic Cells'** are the ones that we are mostly interested in when we talk about batteries.

These Galvanic Cells have the chemicals and elements in them to turn a chemical reaction into DC electricity. So here, unlike the electrolytic cell, the **anode** would be **Negative (-)** and the **cathode** would be **Positive (+)**. They produce DC electric power in their own right and do not need the help of any external power sources.

What is a Galvanic Battery?

This Galvanic battery is what the book is about! A **Galvanic or Voltaic Battery** is a device which turns energy from a **Chemical Reaction** into **DC Electrical Energy** and this would be called 'Discharge' and in many cases we can reverse that chemical reaction back into chemical energy and this would be called '**Charge**'. Some batteries complete their discharge and are what we term as 'flat' and that is the end of their life and must be disposed of.

Batteries come in many different types and in all shapes and sizes. As we continue through the book we will discuss the different types and sizes and how the battery actually works. Let us look at the parts that make up our galvanic battery:

The Case:

The purpose of the case is quite obvious, it holds the electrolyte and electrodes which we will discuss later. We start off with the battery case which relies mainly on the type of battery. If we firstly look at several different batteries that you may be familiar with: the **Cylinder Battery** which you may recognise by a triple 'A' (AAA), double 'A' (AA), a 'C' type battery in many toys, a 'D' type battery that you put into torches and the PP3 batteries that go into your smoke alarms.

There are many more of the above types which are not so common. The case of the most common cylinder battery can be made of plastic-coated **Zinc** if it is, say, a Ni Cd (Nickel Cadmium) or a Zinc-Carbon battery. In alkaline cylinder batteries the case may be made of '**Steel**'. The next set of batteries to look at are the larger batteries similar to the ones used on uninterruptable power supplies (UPS) and the type that you may have in your car. Now the case may be made of **Plastic, Steel, Stainless Steel, Glass and Aluminium**.

The Electrolyte:

Before we can even start discussing batteries in detail we need to know what one of the most important materials is which makes up a cell, and that is the **Electrolyte**. What is this material used in batteries and how does it fulfil its goal?

An electrolyte is a material that when mixed with a **'polar' solution** i.e.: distilled water, produces a liquid that will electrically conduct i.e.: saline solution, acid, alkaline to name three liquids and in some cases gases can be electrolytic. The solution is made up of around 35% acid and 65% water (this % may differ with types of battery!) This may be called '**Aquatic Electrolyte**'.

So when we carry out battery maintenance, what checks and tests must we carry out on the electrolyte? Well firstly some batteries are sealed (Maintenance Free) so here levels are fixed and sealed inside of our tank which cannot be opened, but if the tank is not sealed we must check the electrolyte level in the battery. Let us take a common acid electrolyte – the acid can be diluted by distilled water which can evaporate. **Acid DOES NOT evaporate but it DOES mist!**

Sometimes, in not so common circumstances, acid has to be added if, say, the battery electrolyte spilled out, to maintain the specific gravity (SG). This operation must **NOT** be completed without advice from manufacturers. Acid quantity should remain about 35% of the solution. Distilled or even de-ionised water has **not got** any impurities unlike tap water which could reduce the life of the battery. DO NOT USE ANYTHING ELSE BUT DISTILLED OR DE-IONISED WATER! Some batteries are shipped dry so full electrolyte should be added as per manufacturer's instructions.

The acid electrolyte has a certain density (SG) which must be maintained and checked with a hydrometer to manufacturer's recommendations. In a sulphuric acid battery for instance, if the level of electrolyte drops too low and exposes the plates, a phenomenon called 'Sulphation' takes place where that part of the plate that is exposed becomes covered in crystals and may stop working and reduce the life and efficiency of the battery. If treated quickly this can be reversed and the sulphur taken back into the electrolyte. (See the SG section.)

The Battery Electrodes:

Another pair of the most important parts which make up a battery are the electrodes i.e.: Positive (cathode) and Negative (anode). These with the electrolyte are going to provide the chemical reaction required to make the DC electricity. So what are the materials used for electrodes in batteries and how do they fulfil their goal? Some of the materials on the **Positive (cathode)** are Lithium, Nickel, Hydroxides, Lead, Zinc and Brass and on the **Negative (anode) Carbon**, Graphite, Hydrogen, absorbing alloys etc. Research into polymer cathodes is being carried out.

Sum Up

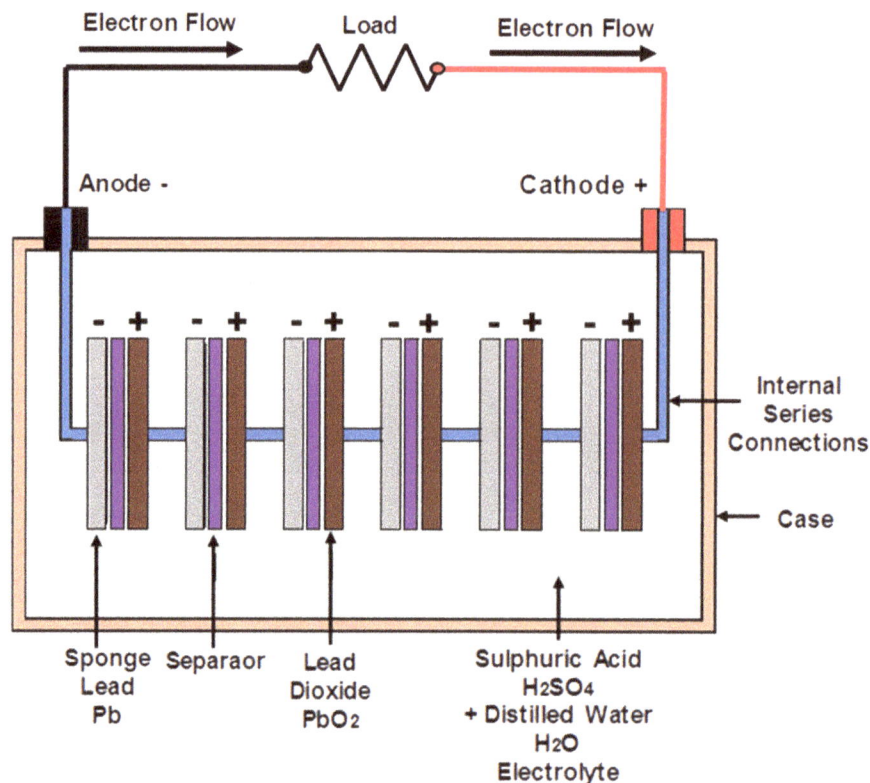

Above is a diagram of a Lead Acid battery showing the parts that we have discussed and a load connected up to it.

To sum up, we have the negative ions collecting on the anode (Negative) and the positive ions collecting on the cathode (Positive) so we now have what is called **Potential Difference** (Voltage) across the battery terminals. Most batteries will remain dormant until a load is connected across them, but some will '**Self Discharge**' meaning they discharge, load or no load.

Now with a load connected across the terminals it will cause Electrons to flow from the anode (Negative) to the cathode (Positive) external to the battery and ions to flow from the cathode (Positive) to the anode (Negative) inside of the battery hence by description these particles seem to flow in opposite directions, but if drawn complete a '**loop**'.

So when the cathode is depleted of particles then the battery is flat and the electrolyte (if liquid) will have changed into a different solution to what it was. In a '**primary**' battery this would be the end of its life, but in a '**secondary**' battery we can put a battery charger onto the terminals and make the particles flow back and change the electrolyte (if liquid) and electrodes back to their original state.

Now you have read through the different battery sections you might now understand more about how they work as we go through the book.

Water for Topping up Batteries:

Wet batteries tend, in the battery world, to be called 'Wet' or 'Flooded' batteries. There are many types of water available, some of which are suitable for topping up **Wet/Flooded** batteries and some which are definitely **NOT** suitable. If you use the incorrect water you could in actual fact shorten the life of your battery quite considerably. Let us have a look at some types of water and discuss which would be the best for our batteries.

Tap Water:

You would think that it would be much handier to top up your wet/flooded battery with tap water rather than distilled or de-ionised water. Let me just say that you will do a lot more harm than good if you were to do this! Why? Firstly normal tap water is a conductor because of the minerals contained within such as Sodium (Na), Calcium (Ca2), Magnesium (Mg2) some of which are **Ionic**, meaning they do not only have **atoms**, but also '**Ions**' so have Electrons to spare so there may be electron flow within the liquid which could counteract the battery's electron flow. Secondly the minerals mentioned could turn to Sulphates, such as Calcium Sulphate, which can coat the plates and make them inefficient.

Rainwater:

Rainwater will gather impurities as it passes through the polluted air so although it might sound ok, it would not be all that wise to use it.

Bottled Water:

Be careful as many bottled waters contain minerals some of which might be similar to tap water. Some minerals are good for us as fit humans, but not so good for our battery. You may be tempted to think bottled water is pure.

Distilled Water:

Distilled water is sometimes called '**Pure**' water and does not contain any of the impurities found in tap water i.e.: Sodium (Na), Calcium (Ca2), Magnesium (Mg2) so there are no minerals to cause 'Sulphation' or ions to cause rogue electron flow. So this water is ideal for wet/flooded batteries. Distilled water is made by a system as you might guess called 'distillation' which is a bit like catching the steam coming out of the spout of a kettle. **Ideal for topping up batteries!** This water is also good for you to drink, possibly chilled, but may taste bland compared to spring water or tap water.

De-Ionised Water:

This water is completely different from distilled water. Often referred to as demineralised water. (Used in large industrial boilers.) The water is passed through electrically charged cation & anion resin beds to remove its ions which tap water would be full of. As it passes through the resin filter, impurities are also attracted out of the water as they are attracted to the opposite charges.

It is reportedly purer than distilled water. As above, companies with large industrial boilers will use de-ionised water as it is anti-clogging. Ideal for topping up batteries! Water filters in your home use '**Osmosis**' (a process of diffusion to remove impurities) to create de-ionised antioxidant water to drink. This water is good for you to drink, possibly chilled, but may taste bland compared to spring water or tap water.

Overfilling:

It is important not to overfill your battery because you can alter the specific gravity (SG) and spillage out of the top would be corrosive. Never of course add acid to top up without advice.

What are Amp hours (Ah) and Watt hours (Wh):

Amp hours (Ah)

We talked earlier about exactly what an electric current was and how we measure it in amperes (Amps). Well if our battery was a 4Ah (Amp hour) battery and the equipment that it feeds takes 4 amps then, when switched on, the power would last 1 hour. The battery, if secondary, would then have to undergo a **'Charge'** which is mentioned later.

Battery Amp hours (Ah) are found by multiplying how many amps the battery is capable of supplying **(Amps)** by the discharge time **(the time it takes the battery to go flat at maximum output in hours)**. So if the battery's output can supply a current of 10 amps for 20 hours then it is a 200Ah (Amp hour) battery (10 x 20 = 200). This can be called **'Battery Run Time'**.

If batteries are purchased, for example, for our chemical plant emergency back-up system or UPS (**U**ninterrupted **P**ower **S**upply) they will be in what is called a **'Bank'** and should be well overrated as far as Amp hours are concerned to enable the lights to be on for as long as possible. Remember that with a UPS system there is usually a piece of equipment called an Inverter which changes the battery DC (Direct Current) to AC (Alternating Current). Changing AC to DC is easy. It is done in the case of your home garage battery charger, but changing from DC to AC is a little bit more complex as you will find out later in the book.

Can I connect batteries with different voltages as per diagram, but the same Amp hours in series? The answer is **YES** so the voltages would be added together 12V + 6V = 18V, but the Amp hours would remain at 4.5.

Be careful connecting batteries of different 'Ah' in series! What you have actually done in the diagram on the right is double the voltage 12V + 12V = 24, but the Amp hours are only **4.5Ah** as per the weakest battery **NOT 8Ah.**

What you **CANNOT DO** is connect two different **VOLTAGE** batteries in parallel in a circuit as all that will happen here is the larger voltage battery will all the time be trying to charge the smaller voltage battery to create a balanced circuit which cannot be achieved here! But what happens in the case of different Amp hour batteries in parallel? Try your best to ensure that the batteries are **IDENTICAL** in make, voltage, Amp hours, charge time, sealed/not sealed etc. You may end up with different charge times etc. The output for the batteries in the diagram to the right would be 12 volts (2 x 12V batteries in parallel) and **12.5Ah.** If you are designing the circuit yourself it is always wise to check with battery manufacturers. One battery may go flat quicker than the other! Completely flat may damage batteries!

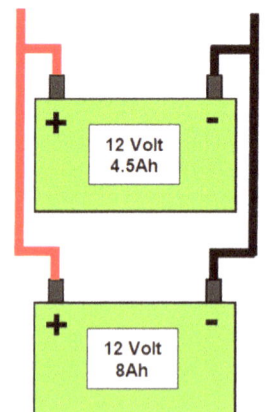

Some battery banks/systems have batteries that are connected in both parallel and series together. If this is the case serious calculations and compatibility **MUST** be looked into if you are designing the system yourself. There may be serious consequences for miscalculation such as output, charge times, and of course Amp hours (Ah). Expert advice is warranted here.

Watt hours (Wh):

We talked above about Amp hours (Ah). Well we can have units of measurement in **Watt hours (Wh)** so 1Wh would mean that the battery will give: 0.5 watts for 2 hours, 1 watt for 1 hour or 2 watts for 0.5 hour etc. Watts are units of work rather than current.

Battery Voltages:

We have looked at current (Amps), now let us look at voltage (V). Let me start off with an analogy: if our electric cable was water pipes full of circulating water, the battery would be the pump, the water would be the current and the pressure of the water would be the voltage. So let us say that **Voltage is Electrical Pressure.** Two things are measured in volts, namely:- **Potential Difference (PD)** and **Electromotive Force (EMF)** although EMF and voltage are two slightly different things. Let me explain below:

Potential Difference:

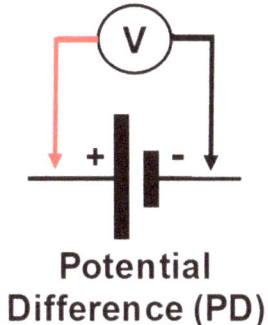

Potential Difference (PD)

The diagram on the left shows a battery with its correct 2 line symbol which has been used since the invention of the **Voltaic Pile**. If you looked at the Voltaic Pile it has long and short plates resembling the symbol. Alessandro Volta (1745–1827) an Italian Physicist invented the Voltaic Pile in 1799.

We use a voltmeter, which is explained later in the book, to measure the voltage or potential difference (PD) across two points, or in our case on the left of the battery. It could be across a resistor or lamp in a circuit. **One volt is defined as one joule of energy per coulomb of charge.**

Electromotive Force:

To make the electrons and hence the current flow around a circuit we need something to 'push' it and that something is a **voltage.** This is slightly different from electromotive force or EMF! So **EMF (E)** is produced by the chemical reaction in the battery and **voltage (V)** is the result of that production! I have shown voltage from positive to negative, many books have this the other way round. **Does it actually matter? NO!**

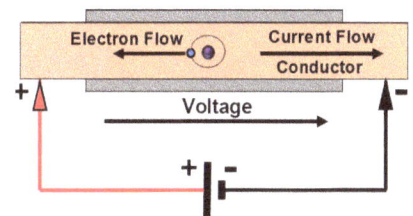

Battery Potential Difference (V)

I apologise for the confusion of having electron flow the opposite way in the circuit to current flow, but remember this is just a theory and it depends upon which book or which site you look at.

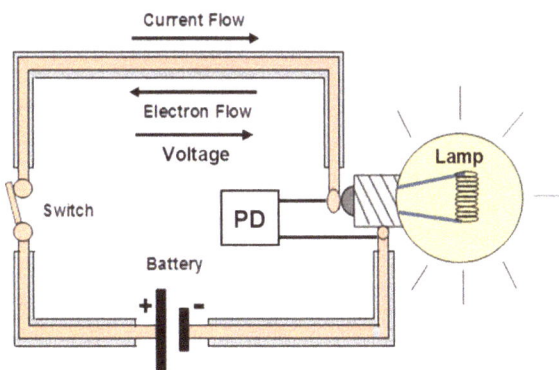

So let us sum this up. The battery is a chemical reaction capable of producing a **PD** and hence an **EMF**. The circuit is connected to the lamp and the switch turned on. Electrons flow from negative to positive and the current and hence voltage in our case from positive to negative making the lamp light! The potential difference is across the battery, or in our case in the diagram left, the lamp.

Other devices besides batteries can also produce an EMF such as generators, fuel cells and solar cells etc. I hope this has not been too confusing!

Open Circuit Voltage:

OCV will always be higher than the circuit voltage under load. This is the potential difference across the battery with everything disconnected. If the battery is supposed to be 12 volts, as a car battery, when you measured the open circuit voltage it may be around 13.5–14 volts. If the open circuit volts of a 12-volt fully charged battery was only, say, 10 volts then the battery may have internal problems or is almost completely flat or could have Sulphation problems (explained later).

Resistance:

Insulation   Resistance (Ω)   Insulation

Conductor   Conductor

Insulation   Insulation

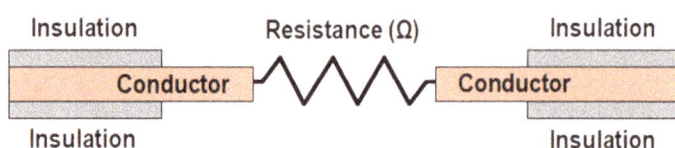

We can say that **Resistance** is the opposition to a current. It is measured in Ohms (Ω) and we can thank Georg Ohm (1784–1854) a German Physicist for its discovery.

From this man we also obtained Ohm's Law (V = IR) which we will discuss later.

The resistor requires a potential difference (PD) to allow the current to pass through it. Looking at the above diagram we now have three mediums for the current to travel through. **Conductors** are where the current, hopefully, will easily pass along, **insulators** where the current, again hopefully, will not travel and a **resistor** where the current will travel very slowly. The slow travel can be put down to a much denser system of Electrons as well as voltage drop.

We have discussed conductors such as Silver, Copper and Aluminium and how the current travels along them easily by movement of Electrons. We discussed insulators such as PVC, rubber and plastic, which keep their Electrons in very tight orbits so will not flow. Now we come to resistors mainly in electrical equipment - toasters, kettles, hairdryers etc. - with elements which may be made of materials such as **Tungsten, Nichrome (Nickel Chrome), Kanthal & Cupronickel.**

Let us say that the current is moving through the conductor satisfactorily and comes up against a resistor. The moving charged particles that make up the current have a certain energy and when they come up against a resistor in the form of fixed particles, some of that energy is lost through **heat**, so if you like the moving current particles may lose some of their momentum. So by using Ohm's Law we can now see that if the value of resistance alters, then other values such as voltage and current may be changed (V=IR).

So we associate resistance with temperature and sometimes that temperature is handy and sometimes it is a nuisance. I will just briefly mention how we can use temperature another way and make a **'Superconductor'** here. Instead of the current causing temperature, if we purposely drop the temperature of certain conductors their resistance starts to decrease rapidly. The resistance of certain materials, if they are cooled to cryogenic temperatures, ceases to exist. This temperature is around -253 degrees centigrade and is called the **'Transition Temperature'** and we have now got our 'Superconductor'.

If we look at the above diagram as a water system where the conductors were pipes and the current was water, voltage would be the water pressure. So there is a volt drop at either end of the resistor or, in our water system, the water pipe then narrows so two things could happen – either the water slows down or the pressure changes, thus in either case energy is lost. We can use the energy, in the form of heat, lost from the current passing through the resistor in devices such as light bulbs or lamps where it is turned into light. We can use the energy in the form of heat from an electric fire.

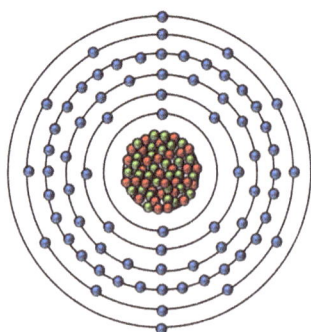

The diagram left is of an atom of **Tungsten** which is an example of a commonly used resistance, but a good conductor of electricity. Its periodic symbol is 'W' & atomic number 74 which means it has 74 Electrons & 74 Protons. Its Neutrons number 110. Used in old light bulbs. Tungsten resistance is around twice that of Aluminium! Looking at older lamp elements, the Tungsten had to be coiled and then the coil coiled again to get the amount of material because of course you cannot join it inside of the light bulb as its temperature could reach a huge temperature of around 3000 degrees Celsius. See how dense electrons are?

Batteries have an internal resistance of around 0.1–2Ω depending upon the type of battery. The conductors, electrolyte etc. all are sometimes not perfect conductors and go to form this internal resistance. There are two types of battery internal resistance, **'electrical'** and **'Ionic'** which is called the **'Total Effective Resistance'**.

Shelf Life & Storage:

Dry Batteries:

Most batteries will stop reacting when disconnected in storage, but some will continue with a slight reaction and this is where the shelf life comes in. The longer the battery is stored the more the reaction and flattening.

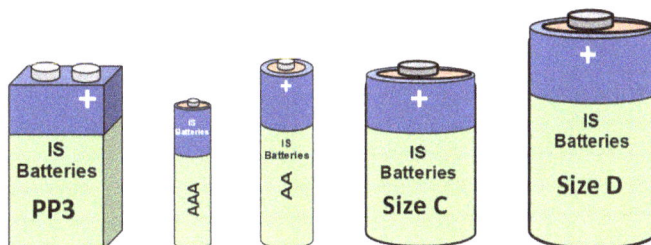

Let us look at battery **'Shelf Life'** and the five examples above are a good starting point. We talk about a battery being a **Chemical Reaction turned into DC Electricity.** We said that when the chemical reaction has fully completed and the electrolyte and electrode chemicals have fully changed to different chemicals the battery is **'Flat'**.

Primary Batteries:

If the battery is flat and it is a Primary or Dry Cylinder Battery then that is the end of its life as there is NO WAY of reversing the chemical reaction. IT WOULD BE DANGEROUS TO TRY AND RECHARGE. THE BATTERY COULD SPLIT OR AT WORST EXPLODE!

Secondary Batteries:

If the battery is a **Secondary Cylinder Battery** then by using a **Battery Charger** across the electrodes we can to a large extent reverse the chemical reaction back to what it was when the battery was new so now it would be recharged. In many batteries this chemical reaction slows and only happens fully when the battery is connected up to an appliance that is **SWITCHED ON**, say a torch. When the torch is **SWITCHED OFF** or the battery is not in any appliance, but stood on a shelf, **in theory** this chemical reaction stops depending upon the type of battery. Many batteries however keep slightly reacting even when not in use, hence the shelf life.

Temperature:

The nearer to **Zero Degrees Centigrade** that the cylinder battery is kept at when not in use, the longer the shelf life span may be. Anything over **Manufacturers maximum storage temperature recommendations** will cause certain internal chemicals to deteriorate and thus start to reduce the battery life. Also I mentioned above **"In theory this reaction stops or slows"**. In practice in some batteries even if the circuit is off or the battery is stood on the shelf then there is still a slight reaction which again may increase with **Temperature**. If the battery is a cylinder type **Primary Battery** you cannot do too much to reduce the loss of charge. If the battery is stood for very long lengths at a time, especially at a higher temperature than recommended, then the chemicals and elements that make up the electrolyte and electrodes will degrade and the life will be dramatically cut down and the battery will not last as long as it should.

If, however, the battery is a **Secondary Battery** and it is stood for long lengths of time, especially at a higher temperature, the chemicals and elements that make up the electrolyte and electrodes could degrade the same as a primary battery and consequently the battery may not take up some of its recharge. **This is called 'Ageing'** and hence the life will not last as long and this of course, as we keep saying, in a primary battery, cannot be reversed.

Periodic Recharge:

The manufacturers of the rechargeable cylinder battery may offer a solution to shelf life - they may recommend that every so often the rechargeable batteries are given a full or partial charge for a set time period. Also the manufacturers may recommend that the rechargeable battery is flattened completely and then recharged now and again, but **BE CAREFUL, NOT EVERY BATTERY!**

There are certain batteries that if they are discharged for storage they must not under any circumstances have their voltage dropped below a certain value. Lithium batteries spring to mind where they must not be allowed to drop below **2 volts** and manufacturers may advise periodic 'top up' charges to prevent this happening.

Larger Wet (Flooded)/Solid Electrolyte Batteries:

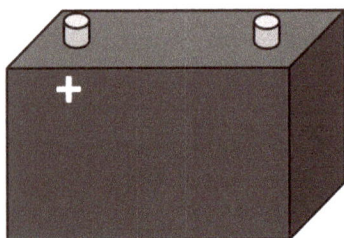

Batteries are devices that change a chemical reaction into DC electricity. I get constantly asked "what is the difference between a battery and a cell?" Well a cell is only one unit whereas a battery is made up of multiple cells or units such as the diagram left. The battery types here are larger type batteries for instance that you may have in your car or UPS systems! Usually, unless sealed, much more maintenance is required for these larger batteries.

A typical battery consists of a **case** which can be, for instance, **plastic.** Inside of this case is what is called an **electrolyte**, which can be many different chemical substances in either liquid such as Sulphuric Acid (H_2SO_4) in, say, the **Lead Acid Battery** or solid in, say, a **Lithium Ion Battery**. The anode (Negative) and cathode (Positive) plates can be intermixed inside the electrolyte forming **'Cells'** with what is called a **'Separator'** dividing them to stop the anode and cathode from touching. The **'Plates'** can be made of, for instance, Zinc, Lead, Lithium etc.

Sealed/Maintenance Free Batteries:

This type of battery can be sealed or have individual chambers where they must be kept topped up with distilled or de-ionised water. Now storing a sealed battery is easier, because due to their make up, the chemicals that go to make up the electrolyte and electrodes have a fairly long shelf life and do not degrade too much and can be put down mostly to three things, namely **'Charge' 'Age' and 'Temperature'**. Manufacturer's advice on charging the batteries during storage must be followed to the letter, for instance they may say that a 'top up' charge must be completed every so many months, say **6 months**, and they may recommend a storage temperature of, let us say, **20 degrees centigrade**. Providing manufacturer's instructions are followed and the sealed battery is stored at the correct temperature, the self-discharge will be around, say, **3% per month**.

If a sealed battery, especially Lead Acid, is allowed to discharge towards flat for long periods then 'Sulphation' of the plates will happen causing them to break down, and no amount of charge will bring the battery back. Now some manufacturers may recommend that their particular battery must be stored with the charge **'Flat'** or a certain percentage below full charge.

Storing with the battery 'flat' may not be so common and many batteries will suffer if you try to store them this way. Lithium batteries for instance may have a recommendation that they are not stored at full charge otherwise they may become unstable. A Lithium battery voltage must be checked as per manufacturer's instructions and must not drop below a certain level i.e.: 2 volts as irreparable damage may result.

We said in the case of a cylinder type battery that the nearer to zero degrees centigrade the better, but in this case it must be remembered that the battery may have a water-based (Aqueous) electrolyte which may not react well to freezing conditions. On the other hand, higher temperatures will cause the battery self-discharge to increase.

Specific Gravity:

When you measure **Specific Gravity**, carried out with a **Hydrometer**, you are measuring the amount of acid and water in the electrolyte and how far on the chemical reaction of a breathable wet/flooded battery is. First of all, to find the specific gravity (SG) you must first have to know either the kg/cm³ or g/cm³. Once you know either of the above values you divide by 1000 if the spec is kg/cm³ or divide by 1 if the spec is in g/cm³. So if a solution was 90 kg/cm³ then the SG would be 90 divided by 1000 = 0.09 so it would float on the top of the water and if the solution was 90g/cm³ then the SG would be 90 divided by 1 so it would sink under the water.

Let us look at what readings of SG you can maybe expect:

Charge:	SG:
100%	1.265
75%	1.225
50%	1.19
25%	1.155
FLAT	1.12

The table to the left shows what hydrometer readings you should expect at various stages of a battery from fully charged to flat. So this is how high the float is in the solution inside of the hydrometer. Remember that readings should be taken after the battery has been on a short load not when it has been stood for some time as the acid may be more at the bottom and water towards the top. I have put a diagram of a hydrometer below. Insert the tube into the battery electrolyte and squeeze the bulb and release which will suck some electrolyte into the chamber and the float will indicate SG.

Looking at the table to the right you can see that water = SG of 1 and acid = SG of 1.052 so the acid will automatically sink below the water which is why the battery must do some work before the SG is taken to mix the two.

ACID	1.052
WATER	1

Hydrometer & Float Position:

It is possible to get electronic hydrometers with digital displays capable of direct SG read out. The older version is the one below.

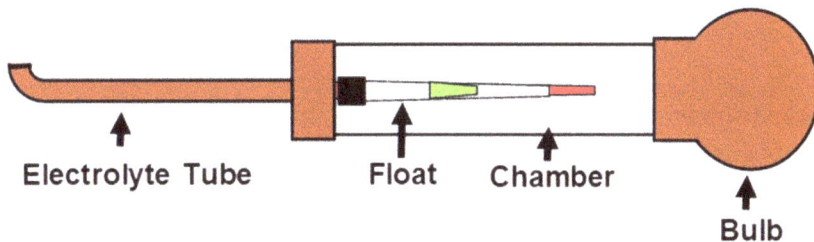

Electrolyte Tube   Float   Chamber   Bulb

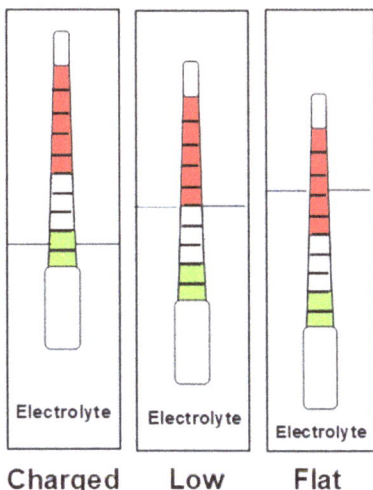

Electrolyte   Electrolyte   Electrolyte

Charged   Low   Flat

Firstly to use the hydrometer (above) put on your eye protection and gloves. Make a note of the location of saline solution bottles.

Secondly insert the electrolyte tube into the battery and squeeze the bulb at the top and allow electrolyte into the glass chamber, squeeze out and re do. There should be enough so that the float rises in the fluid. **Do not carry out when the battery has just been topped up with water.**

Look at the float position in the chamber around half to three quarters full of electrolyte. The diagram on the left shows the float position for charged, low and flat battery. There will be numbers on the red white and green sections of the float corresponding with the table at the top.

This test is usually carried out on wet/flooded **acid** batteries to test the specific gravity of Sulphuric Acid as alkaline batteries do not vary very much at all in specific gravity from charged to discharged, and may only alter as the battery is coming to the end of its life.

Battery Charging:

General:

Battery charging can be a very tricky business and should always be carried out to manufacturer's instructions. Several questions might be: if I keep partially charging the battery can I incur damage by overcharging? Can I undercharge the battery? What if a wet/flooded battery is allowed to stay flat for long periods of time, can the plates Sulphate? If secondary dry batteries are allowed to stay flat for a long time, why do the contents leak out forming a corrosive white powder? Batteries charge because the charger puts a certain voltage/current onto the electrodes. With a vehicle this charge is constant from the alternator. The voltage has to be slightly more than the actual battery output and the current is adjusted accordingly. How high can this external voltage be? Well for a standard secondary 12-volt battery the charging voltage would be around 13.5–14 volts. The current is adjusted accordingly determined by the charge. On the following three pages on charging are several terms that may answer a few questions.

Absorption Charge:

If a battery is completely flat a long charge may be required to bring it up to full charge. When 80% of the battery charge capacity is reached at the rate of charge this is called the **Absorption Charge** and the current should taper off, which depending upon the battery and how flat it is, then the charge to this level could take several hours.

Boost Charge:

Although manufacturers may say this is ok for things like smartphones and other electronic devices, it may not be ok for other types of battery. If we take a wet/flooded battery as in your car, here is where a lot of damage can be done if it is boost charged too often.

We all know the situation, where you have maybe left your lights on, and you are going somewhere in the car. You turn the key and the engine makes that horrible growling sound when the battery is flat. You rush and get your charger and put it onto **Boost** to charge the battery up fast. So by putting a boost charge onto the battery, this will be a high current charge. Doing this too often can cause heat or the plates to buckle and plate material fall off making the battery less efficient. Material/sediment in the base can actually cause the plates to short out.

Charge Acceptance Current:

The amount of charge that a battery can accept in normal conditions in a set time.

Charge Efficiency (%):

This is to do with how much energy was removed from the battery when it was working (discharging) compared with how much energy is used to bring the battery to full charge. Sometimes referred to as '**Charge Acceptance**'.

Charging Current:

The '**Charging Current**' of a battery should be 10% of the amp hour (Ah) rating so a 150Ah battery charging current would be 150 x 10(%) over 100(%) which would equal 15 amps.

Charging Rate:

When we look at the '**Charging Rate**', this is called the **'C' Rate** which is the time it takes the battery to charge i.e.: 2C is 30 minutes, 1C is 1 hour, 0.5C is 2 hours etc. How do we find this 'C' Rate? Well on small secondary batteries the 'C' rate should be either on the side of the battery or on the datasheet. Not all types of battery will be the same. On larger batteries the **'C' Rate** is worked out using the amp hour (Ah) rating of the battery. So to work 'C' out:- t (time) = 1 (hour) divided by the 'C' rate. The answer will be in hours.

Dry Charged:

If a flooded battery is '**Dry Charged**' then it is charged without any electrolyte added. This is probably done by the manufacturer and is advantageous if transporting or storing the battery. This is done of course with an unsealed battery, for example of the lead acid type used on a car. I must say at this point that the majority of alkaline or acid batteries will be charged '**Wet/flooded**' meaning the electrolyte is already added.

Equalisation Charge:

Used on, say, a UPS Battery Bank after a number of cycles or a time period to bring all of the batteries up to the same level of 100% charged.

Equalisation Mode Charge:

This phase is intended to remove the 'Sulphation' off the plates in, say, a lead acid battery. A small charge is put onto the battery for several hours to do this, with electrolyte topped up.

Float Charge:

The Float Charge is usually on, say, back up secondary batteries which may not be used that often. The current is monitored by the charger and just keeps the battery topped up and adjusts or switches off when the battery charge is full. Smart chargers will accomplish this procedure very efficiently. This may be called a **Constant Current Mode/Charge**. Not the same as '**Trickle**' Charge.

Formation:

Formation is charging the battery for the very first time. (Manufacturing).

Gassing:

Caused, for example, by overcharging where the cells release Hydrogen & Oxygen gas.

Inductive Charge:

Uses Electromagnetic Induction to charge portable equipment such as smart phones instead of physical leads. Just mount the appliance near the charger.

Memory Effect:

A minority of older batteries have a **Memory Effect** which means that if you were to keep charging them before their charge was depleted, a 'Memory' would build itself into the battery where it would only accept a percentage of full charge. **Nickel Cadmium** springs to mind.

Minimum Charge Voltage:

Is there a minimum charge voltage that I have to put into a battery so that it accepts the charge? Most definitely and that voltage is, for example, over 2.15 volts/cell. Anything less will not charge.

Nominal Voltage:

This is the voltage of the battery when it is half way between being fully charged and very low charge.

Overcharging Nickel Batteries:

If we take the battery used in modern laptops and phones it is very difficult to overcharge but not a good idea to keep them permanently plugged in. **Nickel Cadmium (NiCd)** and **Nickel Metal Hydride (NiMH)** in older devices have a '**memory**' so should be fully discharged before being charged otherwise it will remember and not accept full charge. Lithium Ion batteries on more modern devices do not have a memory and can be charged as often as necessary and cannot be overcharged.

Overcharging Wet/Flooded Batteries:

Other batteries, such as a car wet/flooded acid battery, are a different matter. If we take a lead acid battery, one of the main problems of overcharging is heating up the electrolyte and hence the inside of the battery which in turn will cause the outer case to get hot. The battery can swell or split and at worst case the battery could explode. Once this type of battery chemical reaction cannot accept any further charge then the charge still connected must be dissipated and this is done in the form of heat. Lithium batteries have to have special protection against a phenomenon called '**Thermal Runaway**' causing explosions. So the problems of overcharging go on.

Over Discharge:

Can I over discharge, say, a flooded battery? Yes, it is possible to over discharge a battery and let it go completely flat below manufacturer's tolerance. Damage may occur to the battery if this happens. UPS Systems have a '**Cut Off**' voltage before this happens. If you take your car battery, if the car will not start you will keep trying until the engine will not turn over anymore, but the battery may still have enough energy to light the lights so it is not completely flat.

Pulsed Charge:

A 'Pulsed Charge' requires a special battery charger to give bursts of charge to the battery, say, a few milliseconds apart. The idea is that during charging what we are actually doing is causing the chemical reaction that happened during the battery discharge to be reversed. The charger giving the battery pulses of charge allows the reverse chemical reaction going on inside of the battery to keep up with the charge. If this is not done, a phenomenon called '**Hysteresis**' can occur due to the battery chemical reaction lagging the charge current.

Self-Discharge:

Batteries, as we have discussed, are chemical reactions resulting in DC electricity. Well with most batteries this chemical reaction ceases when the load is removed, but a minority of batteries keep reacting even with no load. This is also to do with **Shelf Life**. Hearing aid Air batteries are a good example of this.

Smart/Intelligent Charger:

Battery charger with **Management System** for charging keeping the charge levels exact.

State of Charge (SoC):

This is a measure of how much battery capacity there is left at any one time.

Trickle Charge:

A **Trickle Charge** and a **Float Charge** are very similar and some people class them as the same thing, but the difference is that once a trickle charger has been connected to a battery it will supply that voltage/current constantly without switching off even when the battery charge is full unlike the float charge.

This may be called a **Constant Voltage Charge/Taper Charge**. If this charge is left on there is a serious risk of overcharging. Trickle charge may be on the battery for around 14 hours at 0.1C or boost charge for say 1 hour at 1C. ('C' meanings are later in this section.)

Wet Charged:

Refers to a wet/flooded battery. If a battery is being '**Wet Charged**' then this is usually the conventional way of charging as the electrolyte has been added and the battery is in use or is ready to be used after storage or transportation.

Primary/Dry Batteries:

Primary/Dry batteries: when they are flat, that is the end of their life and they cannot be recharged **and it would be dangerous to try!** They must be safely disposed of in an environmentally friendly way. The list of primary batteries may be longer than you think.

Examples:

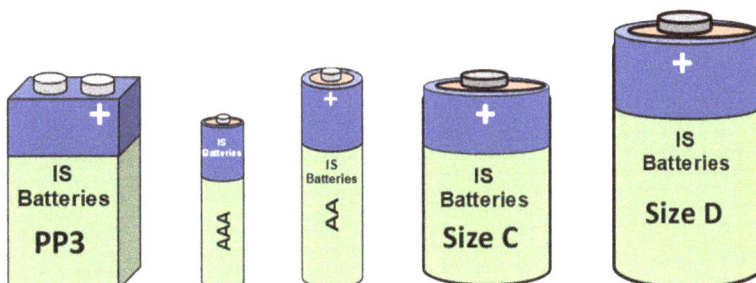

Five of the most common primary battery types are in the diagram above. I must state at this point that **SECONDARY** batteries can be obtained as the examples above **and these can be charged.** See manufacturer's data.

Energy Density:

Primary batteries are high on energy density. This is the amount of energy stored compared with the volume of the battery. All Redox (explained later) effort is put into the output rather than charging. If you look at the chart below at the mAh compared with the size:

Approximate Dimensions & Milliamp Hours

Battery Type:	Length:	Width:	Amp Hours:
PP3	48.5mm	26.5x17.5mm	570mAh
AAA	44mm	10mm	1,000mAh
AA	49mm	14mm	2,500mAh
C	49.5mm	25mm	7,000mAh
D	59mm	33mm	15,000mAh

The table above is just to give you an idea of the dimensions of the battery compared with the approximate milliamp hours of the batteries at the top. So the PP3 which you would install into your smoke detector would provide 570milliamp hours which is 570milliamps for 1 hour. This is longer than it sounds as these appliances take fractional current. These batteries are not designed for constant high current use.

Safety – Carrying batteries in pockets:

Never carry batteries openly in your pocket, especially in a Hazardous Area. If we take the PP3 battery, if the positive and negative terminals were to come into contact with, say, a set of keys in your pocket and shorted out, the heat generated inside of the battery can cause it, in extreme circumstances, to explode or catch fire. Carrying open batteries is like carrying a box of matches!

It is almost certain that throwing batteries onto a fire or artificially heating them will cause them to rupture and explode. They can fly out of the fire like a bullet and cause severe injury if someone is unlucky enough to be in the way.

Primary Batteries work on the basis that once the chemical reaction inside has taken place then the battery is **FLAT** and its life is over and there is no way to reverse the reaction to charge it back up. A typical example of a more common primary battery is the **Zinc-Carbon Cylinder Battery** that you might buy for a torch. There are many types of primary battery, some quite common and easily recognisable, some not so common. In the following pages I have shown in diagrams and descriptions the most common **Primary Batteries** that you may come across in industry (1–6) and a light description of some other primary batteries that are not so common (7–13 below).

1) Primary Zinc Carbon

2) Primary Alkaline Battery

3) Primary Zinc Silver Oxide

4) Primary Zinc Air Battery

5) Primary Mercury Battery

6) Primary Beta-Voltaic Battery

7) **Primary Aluminium Air Battery:** These batteries are not so common. They work by an Aluminium anode (negative) reacting with an Oxygen (in the air) cathode (positive). Once the Aluminium is depleted the battery is flat and that is the end of its life. I think there is potential in the future for Aluminium batteries because of its light weight, but much more research needs to be done.

8) **Primary Atomic Battery:** These batteries may be called **Radioisotope Batteries.** I don't think that I have to describe the hazards if we were to use them daily. They use the decaying isotope reaction to generate electricity just like a chemical reaction. They are used where there may not be much attendance required. Uses i.e.: space, heart pacemakers etc. These batteries are identified by the radiation logo on the side of the battery. Hazards also apply in manufacture.

9) **Primary Lithium Battery:** These batteries are usually called Lithium 'Metal' batteries as the anode (negative) is made of Metallic Lithium and the cathode (positive) is Manganese Dioxide with an electrolyte comprising of Lithium Salt mixed in an Organic Solvent (Propylene Carbonate for instance). Again, can be used in heart pacemakers and long life uses. There are of course many secondary Lithium batteries with some that are used in vehicles.

10) **Primary Magnesium Battery:** These batteries have been around for a number of years, but seem to have been overtaken by other primary batteries. These have a cathode (positive) usually of an Alloy of Chlorides (e.g.: Silver, Copper). The anode (negative) is of course Magnesium Alloy. The electrolyte of this primary battery is Magnesium Perchlorate. The voltage of this cell is **2–2.5 volts.** There are also magnesium secondary batteries which are rechargeable.

11) **Primary Nickel Oxyhydroxide Battery:** These primary galvanic batteries are opposite to most with a cathode (positive) of Graphite & Nickel Oxyhydroxide and Manganese Dioxide. Manufacturers use a new electrolyte vacuum pouring technique. The voltage of this battery is around 1.7 volts. These batteries are usually long life.

12) **Primary Water Activated Battery:** These batteries in actual chemicals are similar to the magnesium battery above. Have an anode (negative) made of Magnesium and the cathode (positive) uses various Chlorides. The difference lies in the fact that they are stored dry and are activated at the required time of use by adding water to the electrolyte. Some of these batteries use actual sea water as an electrolyte.

13) **Primary Zinc Chloride Battery:** These batteries are also known as **'Heavy Duty'** batteries. Very similar to the common Zinc Carbon battery but a longer life. The electrolyte is Zinc Chloride with a very small amount of Ammonium Chloride added. Voltage of this battery around 1.5 volts.

Primary Zinc-Carbon Acidic Battery:

So as I keep mentioning that primary batteries that you buy for torches etc. are the ones that are most common, but cannot be recharged. When they are flat that is the end of their life and they should be disposed of in an environmentally friendly way. **NOT IN DUSTBINS!** So let us start with **Zinc-Carbon Acidic Battery:**

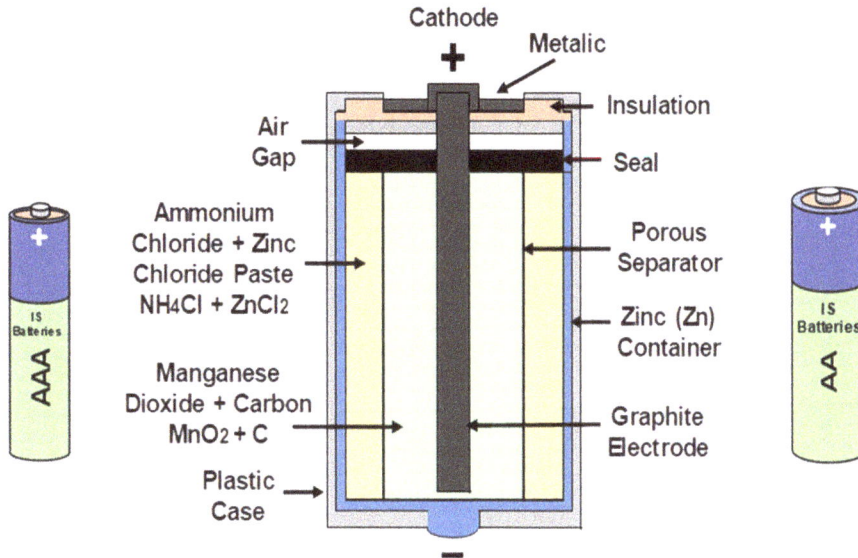

Batteries are devices that change chemical reactions into DC electricity! There are two main types of battery namely: **Primary** batteries and **Secondary** batteries. This example is called a **PRIMARY BATTERY** which means that it has a finite life and when flat that is the end of its life as **IT CANNOT BE RECHARGED AND MAY SPLIT OR EXPLODE IF YOU TRY!**

This common acidic 'cylindrical (because of its shape) battery' is based on a **Leclanché Cell** and shown in the diagram above. This battery is used in many devices such as torches, calculators, children's toys etc. and is made up of a Zinc (Zn) 'enclosure' which in actual fact is the 'anode' (negative) electrode coated on the outside with plastic with the manufacturer's name, voltage and charging details.

Inside the enclosure are two different **'Pastes'** kept apart by a porous **'Separator'**. The pastes are namely: Manganese Dioxide (MnO_2) and Carbon (C) and Ammonium Chloride (NH_4Cl) and Zinc Chloride ($ZnCl_2$) both are corrosive and very toxic. Down the centre is a Graphite/Carbon Rod which is the 'cathode' (positive) electrode. The battery then has various seals and insulators at the top.

The Discharge Formula:- $Zn + 2MnO_2 + 2NH_4Cl = Mn_2O_3 + Zn(NH_3)_2Cl_2 + H_2O$

Most standard primary batteries are **1.5 volts** each. I must warn people that batteries can be obtained of similar sizes that are **1.2 volts** and others that are **1.7 volts** so care must be taken when selecting the type and manufacturer if your appliance is voltage sensitive.

DO NOT MIX BATTERIES OF A DIFFERENT VOLTAGE, MANUFACTURER AND TYPE AS AGAIN THEY CAN SPLIT OR EVEN EXPLODE. JUST IMAGINE THIS HAPPENING IN A SAFETY TORCH IN A ZONED AREA!

Battery:	Voltage:	Use:
AAA	1.5	Games
AA	1.5	Games
C	1.5	Smaller Torches
D	1.5	Larger Torches

The cylindrical batteries shown in the chart on the left are the common primary batteries of the type above that you may come across in everyday life. The chart shows the type, voltage and what they may commonly be used in.

Other Shapes of Acidic Primary Batteries:

(NOTE: Can be obtained as Alkaline Batteries also)

Example: Lantern Battery Type 1:

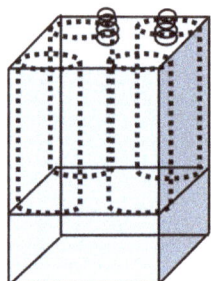

The square **'Lantern' Batteries**, the ones with the two springs on the top as in the diagram on the left, which are used in the red flashing lamps that you see in road works at the side of the road guiding the traffic are **6 volts**, but as previously I have said that a basic cell is **1.5 volts,** how do we arrive at **6**? Well here they simply put a spacer in the base of the square container and install four **'D'** size batteries (standard torch batteries) in series **(4 x 1.5 volts = 6 volts)** inside of the square chamber sitting on the spacer (this type being a primary battery – when it is flat that is the end of its life).

Example: PP3 Battery:

Likewise the **9-volt** PP3 battery that you put in your smoke detector in your house and also many instruments, is simply a stack of **6 x 1.5-volt** Carbon Zinc batteries as per the diagram on the right. Other types of PP batteries can be obtained such as PP1–11 most of which are **9-volt** batteries as well. They are also comprised of **6 x 1.5-volt** batteries in series making the **9 volts** exactly the same as the PP3 diagram to the right. Again what we are talking about at the moment is a **Primary** battery which means that when it is flat it cannot be recharged and is at the end of its life. Secondary **9-volt** rechargeable batteries of the PP3 size can be obtained.

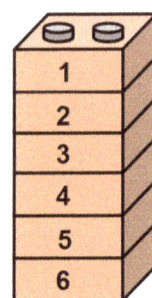

Example: Lantern Battery Type 2:

I am sure that you will have come across this type of battery on the left before. Well this is also sometimes called a **'Lantern'** battery, but is found more abroad than here. This battery has two metal strips on the top and is usually a **4.5 volt** battery so would consist of **3 x 1.5 volt 'D' size** batteries inside. I am sure that by now you are getting the picture of how batteries can give a higher nominal voltage than **1.5 volts.**

Corrosion:

Have you ever taken a battery out of equipment and it is sticky? This is because the Zinc case has been attacked by the Ammonium Chloride inside of the battery and the case has become so thin that this chemical has leaked out. Many batteries of this type have a thin plastic film around the outside to prevent leakage into the air should the Zinc case rupture.

Primary Zinc Chloride:

Just a Note: The Zinc-Chloride batteries are classed as a **'heavy duty'** improvement on the Zinc-Carbon battery. They last longer and are not so fussy on the ambient temperature.

Primary Alkaline Battery:

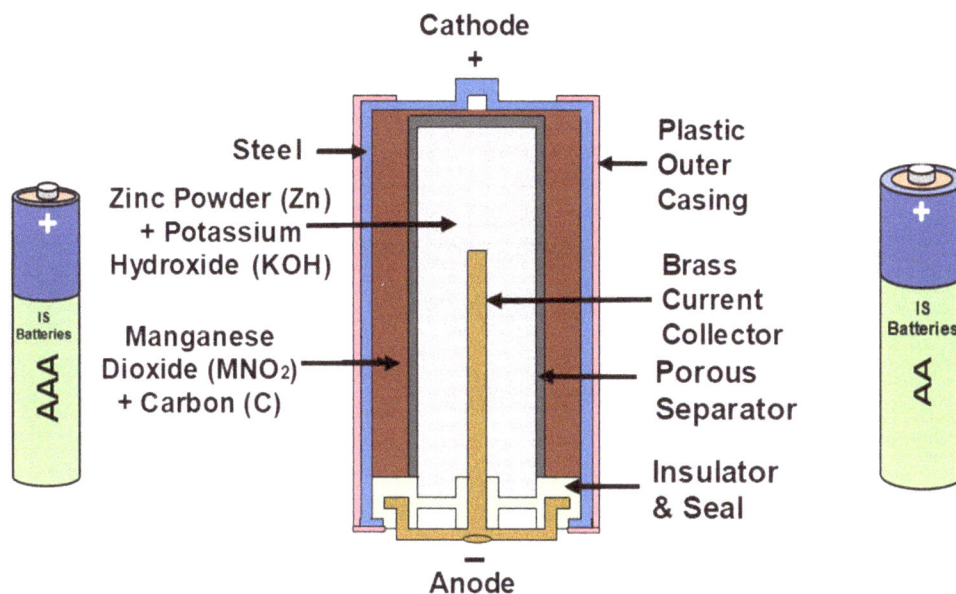

The common **'Alkaline Primary Battery'**, in the diagram above, is used in many devices such as torches and children's toys. The battery is made up of a steel **'Enclosure'** coated on the outside with plastic. Inside the enclosure are two different **'Pastes'** kept apart by a porous separator. These pastes are very toxic.

This is a PRIMARY BATTERY which means it has a finite life and when flat IT CANNOT BE RECHARGED AND MAY SPLIT OR EXPLODE IF YOU TRY! In the centre rising up from the base is a **Brass 'Current Collector' Electrode**. The current electrode is surrounded by a 'Gel' which is a mixture of two chemicals namely: Powdered Zinc (Zn) + Potassium Hydroxide (KOH).

The electrolyte, namely Manganese Dioxide (MnO_2) and Carbon (C), is in a steel outer container which is the cathode. A porous separator separates this electrolyte from the Manganese Dioxide.

DO NOT MIX BATTERIES OF A DIFFERENT VOLTAGE, MANUFACTURER AND TYPE AS AGAIN THEY CAN SPLIT OR EVEN EXPLODE. JUST IMAGINE THIS HAPPENING IN A SAFETY TORCH IN A ZONED AREA!

Corrosion (White Powder):

Sometimes you open the battery chamber of a device which has not been used for some time, especially with alkaline batteries, and find a white powder around the battery. Normally the case will be strong enough to hold the battery gases without rupturing when it is powering equipment. Sometimes, however, in practice, battery chemical reactions continue on a small scale even if the appliance is not switched on, again causing the gas build up.

The gas will build up and rupture the case causing some of the contents **(Electrolyte)** to leak out. The electrolyte may be Potassium Hydroxide (KOH), as in the diagram above, which will react with air to form **Potassium Carbonate Crystals (K_2CO_3) – a white sugary type powder**. This is very corrosive and will corrode any metal springs etc. inside of the chamber. It is also toxic and a hazard to eyes and skin so be very careful about getting into contact with it. Do not let children near it!

If you have battery operated equipment that you know you are not going to use for some time then remove the batteries and prevent the formation of the powder. However if the powder has formed, a small tip may be to **put on eye protection** and use a swab or small stiff brush to gently remove as much as possible from the equipment chamber then keep putting some vinegar or lemon juice on a swab and wipe over the powdered area. This acidic action should neutralise it and maybe save your appliance. **Wash hands thoroughly afterwards.**

Primary Zinc Silver Oxide Battery:

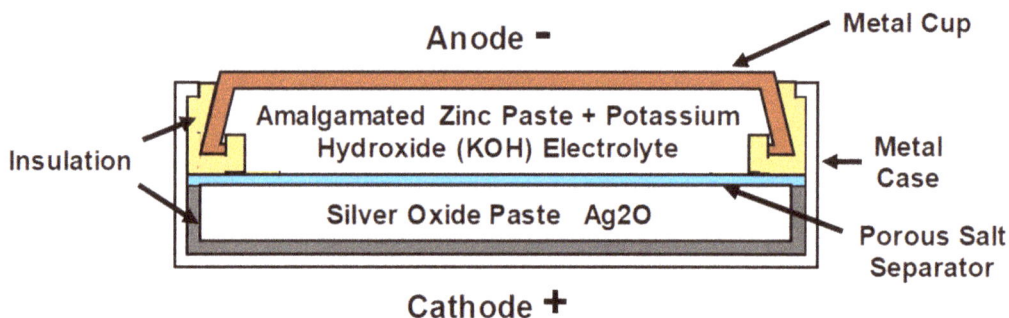

These batteries come in various shapes and sizes, but are very small. They are **PRIMARY BATTERIES** which means that they **CANNOT BE RECHARGED.** Its flat shape sometimes earns it the name of a **'Button'** battery. Used as a power source for equipment such as watches, hearing aids, calculators etc. Although expensive these small batteries tend to give a long life. They have Silver as part of the chemistry which makes them expensive. Usually have a finite life, but it is possible to obtain a rechargeable version of this battery. These may be confined to instances where they are charged by the device, such as a laptop PC, and have to hold information even if the device is switched off or the main device battery has gone flat.

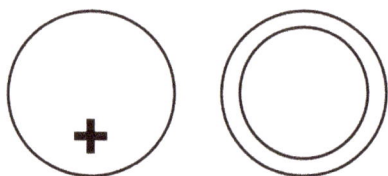

Some of these batteries tend to be made using Silver Oxide (Ag_2O) + Zinc (Zn). Silver Oxide batteries may be expensive because of what it consists of in its chemistry. Other materials can be used in these **'Button'** batteries as below.

At the top of the page is a diagram of a Silver Oxide **'Button'** battery just to give you an idea of the construction, but there are several other similar batteries with different anode materials such as Lithium (Li) instead of Zinc (Zn). Lithium batteries are discussed under their own section in this book. Some different cathode materials can be: Copper (Cupric) Oxide (CuO), Manganese Dioxide (MnO_2) or Carbon Monofluoride (CS) which of course is an alkali.

Something else that can unusually be used as a cathode (positive) material is Oxygen (O_2). Certain hearing aid batteries in the past used to use one material called Mercuric Oxide (HgO) but as you can imagine in an environmentally friendly world anything containing Mercury (Hg) would not be used in modern times. Instead of Mercury (Hg) the manufacturers may use Lead (Pb) which in itself is not the most environmentally friendly element.

The letters which designate the different sizes of these **'Button'** batteries are: C, S, P, L, B, G & Z. The M & N sizes were 'Mercuric Oxide' and are not used anymore. Again the L, S, P, G & Z batteries are the standard 1.5 volts. The C & B sizes are 3 volts so it is very important that the correct size and voltage is chosen.

A typical calculator battery may have the number: **CR2025**. The **first letter** denotes the type i.e.: **C** = Lithium, **S** = Silver Oxide, **P** = Zinc/Oxygen, **L** = Manganese Dioxide, **B** = Lithium Carbon Monofluoride, **G** = Lithium Copper Oxide and **Z** = Nickel Oxyhydroxide. The **second letter** denotes the **shape: R** = Round, **F** = Flat, **S** = Square & **P** = the rest of the shapes not mentioned. The numbers denote the primary element/chemical (i.e.: Lithium), thickness of the battery and the diameter.

So our Battery CR2025 = Lithium Round 20mm Diameter and 2.5mm high.

One of the main safety problems with this type of battery is that it is so small that it can be easily swallowed by children. IF A BATTERY IS SWALLOWED BY A CHILD IT CAN BE FATAL. YOU MUST TAKE TO ACCIDENT AND EMERGENCY IMMEDIATELY! IT MAY LODGE IN THE STOMACH OR INTESTINE AND SOME ARE VERY, VERY DANGEROUSLY CORROSIVE AND/OR TOXIC!

Primary Zinc Air Battery:

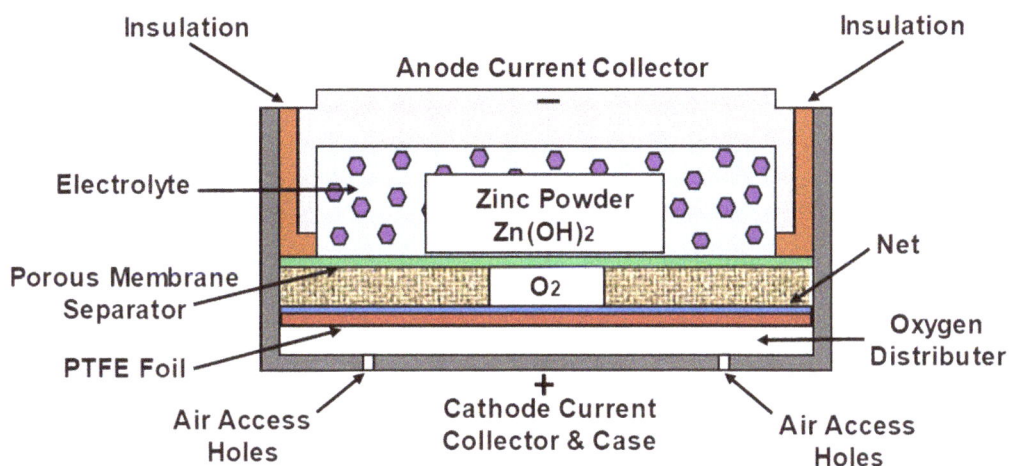

To anyone who ever has worn a hearing aid these batteries will be quite common. Again these are **PRIMARY BATTERIES** which means that they **CANNOT BE RECHARGED.** It is possible to obtain rechargeable Zinc Air Batteries. We will take the hearing aid battery as an example because quite a lot of people use them.

So if you look at your battery your will see that there is a sticker across the **Cathode**. This end is of course the **Positive** and has a large **Positive (+)** sign to inform you of this and you will see that this is the actual container of the battery. This sticker is in place to cover up several **'Air Holes'** which are at the bottom of the diagram above. This stops the reaction taking place and when the sticker is removed you should wait a short while before installing. Next time you pick up one of these batteries look for the air holes, they are very tiny! Once removed the reaction will continue permanently.

The other end of the battery (top in the diagram) is the anode and the small circle on the battery. You will see at the top of the diagram above that the anode is insulated away from the container which is the cathode. If we look at the diagram above you will see that the electrolyte (blue) is at the top and in contact with the anode end of the battery and in our case this is **Granulated/Powdered Zinc.** If you look at the bottom of the diagram above you will see that there are two **'Air Access Holes'**. These allow air into the case and this mixes with the **'Hydroxyl ions'** which then migrate through the membrane and merge with the blue electrolyte in contact with the anode. This process releases electrons which travel back to the cathode.

So removing the plastic tab activates the battery and at that point the voltage of the battery if you measured it would be around **1.1 volts**. After a while, when the air starts seeping into the battery through the air access holes, this voltage should rise to around 1.**45 volts**, hence it is always wise not to put the battery straight into the appliance when you remove the tab but to leave it for a short time. Putting the battery into the appliance straight away will restrict the air holes and hence efficiency.

This battery has a fairly long life, but after the plastic sleeve is removed will continue to work at a low level even if the appliance is switched off.

The Discharge Formula:- $2Zn + O_2 = 2ZnO$

One of the main safety problems with this type of battery is that it is so small that it can be easily swallowed by children. IF A HEARING AID BATTERY SUCH AS THE ONE ABOVE OR THE ONE NEXT IS SWALLOWED BY A CHILD IT CAN BE FATAL. YOU MUST TAKE TO ACCIDENT AND EMERGENCY IMMEDIATELY! IT MAY LODGE IN THE STOMACHE OR INTESTINE AND SOME ARE VERY, VERY DANGEROUSLY CORROSIVE AND/OR TOXIC!

Primary Mercury Battery:

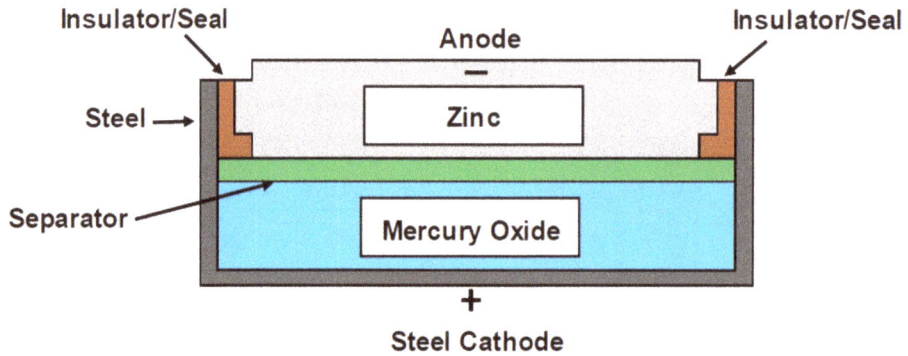

As mentioned earlier, the battery above **WAS** used in hearing aids and watches during the 20ᵗʰ Century. Mercury batteries or cells, as they were called, were renowned for lasting for long periods, but of course they are not environmentally friendly and were replaced by other types of battery such as the Zinc Air battery which we use today. It was the **Battery Management Act in 1996** passed in the United States that actually had these batteries phased out.

These batteries were renowned for holding their charge until the last remaining percentage of their lifetime. The cathode (positive) was an oxide of Mercury mixed with Graphite and the anode (negative) Zinc. The electrolyte was an alkaline of Sodium or Potassium Hydroxide. They also had a long shelf life of years which made them very attractive in their day.

Primary Beta-Voltaic Battery:

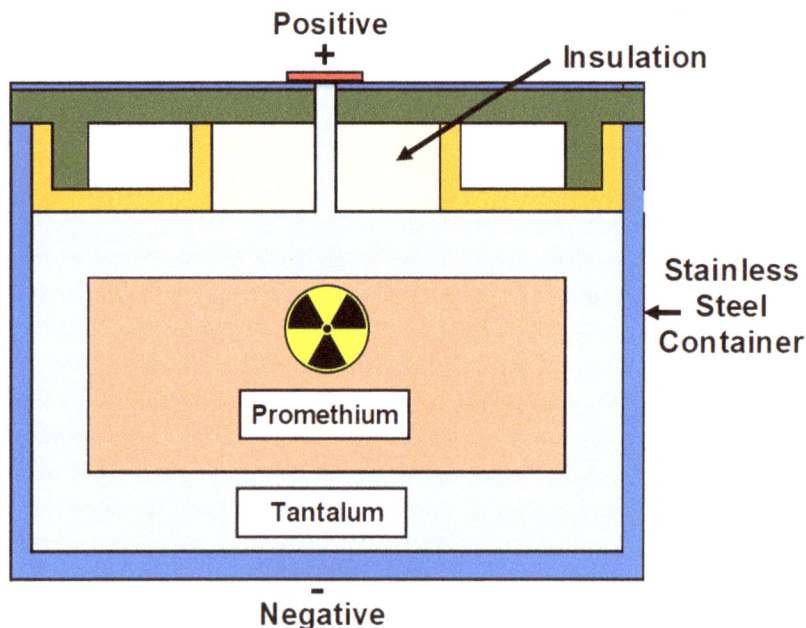

These batteries, although not common, I have put in diagram form just to show what actually happens inside of the battery. Like the nuclear one mentioned earlier, it generates electricity from radioactivity.

They use **Beta Particles**, hence the name. The **'Beta'** particles travel through a semi-conductor leaving behind 'holes'. Electric current is produced by movement of electrons into these holes. This battery has a very long, but finite life.

I do not think that I need to remind anyone about the problems of incorrect disposal of these batteries and the effect they may have on the environment!

Disposal of Primary Batteries:

The big question is why can't we dispose of these batteries in a dustbin? Well we are looking at: the **Human/Animal Toxic Risk, the Environment which includes the Water Table and the Food Chain**. Batteries contain many substances and chemicals which are **TOXIC** and **NOT** environmentally friendly as well as some that are environmentally friendly! What we have to look at is what happens if these materials are non-degradable and get into the **Water Table** or the **Food Chain** through livestock, fish etc. Mercury would be a good example here.

I think it goes without saying but never throw batteries on a fire such as a bonfire as they are likely to explode!

1) **Acid (Electrolyte):** Very toxic and harmful to humans and animals. Will burn on the skin. Don't think I need to go much further except to say **VERY HARMFUL!**

2) **Alkaline (Electrolyte):** Very toxic and harmful to humans and animals. Will burn on the skin. Don't think I need to go much further except to say **VERY HARMFUL!**

3) **Ammonium Chloride:** Very toxic and harmful if it gets into the eyes.

4) **Cadmium:** Toxic to humans, plants & animals. Accumulative and can affect liver, kidneys and bones and is a carcinogen. **Does not break down in landfill!**

5) **Carbon:** Is a Green element unless dust breathed in by humans of course. Many of these substances are found in everyday life but can be dangerous in concentration.

6) **Copper:** Is a Green element. Many of these substances are found in everyday life but like most things can be dangerous in concentration.

7) **Lead:** Toxic to humans at any level. Can affect the nervous system and cause mental difficulty. Accumulative and can store itself in bone. Stops plants' natural mechanisms thriving.

8) **Lithium:** Mining Lithium is of more concern to the environment than Lithium batteries themselves. Toxic if it gets into drinking water.

9) **Manganese:** Toxic to humans and animals in its pure form. Found in almost everything in nature as a percentage, not in pure form.

10) **Mercury:** Very toxic to humans and animals (even though in the 1960s we used to chase it all over the laboratory desks in science!) It affects the immune and nervous systems. Toxicity levels can suddenly go up in soil, animals and fish which of course can be passed to humans through the food chain.

11) **Nuclear Material:** I do not think that I need to put much here for you to understand how dangerous this can be to **ALL** life!

12) **Potassium:** Potassium itself is reportedly environmentally safe. Oxides and monoxides of Potassium may however be another matter. Many of these substances are found in everyday life but, like most things, can be dangerous in concentration.

13) **Zinc:** Zinc is in most natural things including vitamins etc. so is considered to be non-toxic, but like most things in excess can cause symptoms in humans like drowsiness, nausea, vomiting etc.

Recycling Batteries:

What happens to batteries at the end of their life? They have to be recycled as mentioned earlier in this book. Throwing them in dustbins is not an option. Many workplaces have battery disposal systems and many companies run schemes for collecting batteries. When the box is full they will call and pick it up for recycling and leave another.

Secondary Lead Acid Battery:

General:

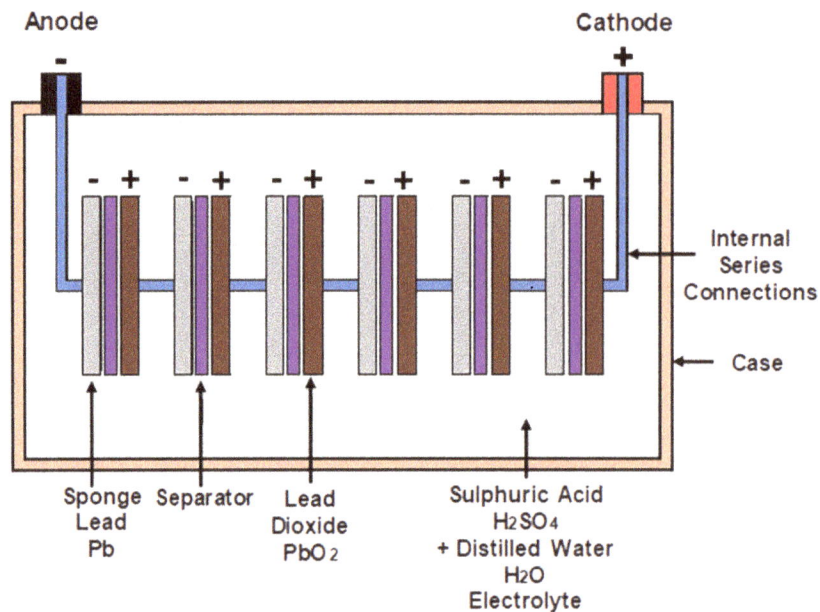

Two types of Lead Acid Battery:

There are two types of Lead Acid battery. The first type is **Starting (Cranking)**. This is probably the most common **Wet/Flooded** secondary battery and you will of course recognise this as the car battery. It will deliver large amounts of energy in one go when required i.e.: when starting an engine. The other type is the **Deep Cycle (Marine)** battery which will deliver energy over a greater time, but not required to start engines. Ideal for electric vehicles such as mobility scooters, golf trolleys etc.

Battery Make up:

As per our book's history section we can thank a guy called **Gaston Planté** for this invention. We have positive plates consisting of Lead Dioxide (PbO2) and negative plates of Sponge Lead (Pb) divided by a porous separator. These are called 'Cells' in an electrolyte mixture of Sulphuric Acid (H2SO4 around 35%) and Distilled Water (H2O around 60%). Liquid electrolyte batteries can be called **'Flooded' Batteries**.

These cells and electrolyte are all in an insulated, acid-resistant enclosure of substances such as glass, resin or rubber composite which of course are insulators. These batteries can be 'Sealed' i.e.: maintenance free **(Valve Regulated Lead Acid – VRLA)** and sometimes have a **Gel Electrolyte (Gel Battery)** or have caps with breathers where they can be topped up with distilled or deionised water. These batteries may be referred to as **Starting, Lighting & Ignition (SLI).**

In the diagram above you will see that there are 6 cells connected in series. Each cell, sometimes called a '**Plate Block'**, is 2 volts so multiplied by 6 would give us 12 volts. I have just drawn one plate for the cathode (+) and one plate for the anode (-) but in reality there are many plates making up each cell.

Plates are used inside of the battery for each anode and cathode with a separator between them instead of a solid block for each to make the battery more efficient by showing more surface area to the electrolyte. The separator is to stop the positive and negative plates from shorting together causing an explosion inside of the battery. Remember that if there is an explosion you have got a battery, in many cases, containing Hydrogen.

Secondary Lead Acid Battery

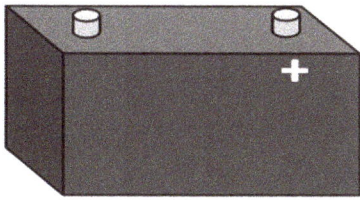

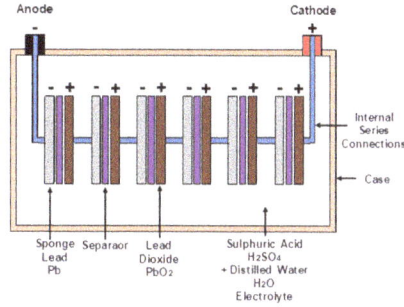

Electrolyte

(Electrolyte H2SO4) When the battery is discharging, only when a load is attached, the electrolyte materials/chemicals slowly change to a different material/chemical and in doing so give a voltage at the battery terminals and when this chemical reaction is fully complete the battery is flat.

Note: I keep reminding readers that with a primary dry battery its life would end here as there would be no way to change the reaction back.

The chemical formula for this Lead Acid Secondary battery is:

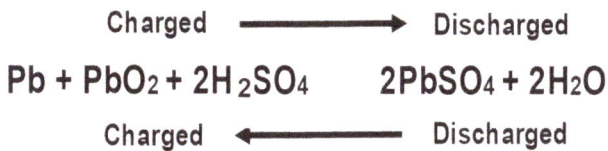

$$\text{Charged} \longrightarrow \text{Discharged}$$

$$\textbf{Pb + PbO}_2\textbf{ + 2H}_2\textbf{SO}_4 \qquad \textbf{2PbSO}_4\textbf{ + 2H}_2\textbf{O}$$

$$\text{Charged} \longleftarrow \text{Discharged}$$

To charge the flat battery we now have to change the 2PbSO4 **(Lead Sulphate)** + 2H2O **(Water)** back to what it was to start with i.e.: Pb **(Lead)** + PbO2 **(Lead Dioxide)** + 2H2SO4. **(Sulphuric Acid).**

Charging:

To do this we put a voltage onto the battery that was greater than its output voltage. This would be called a **'Trickle'** charge. If we increase that charge current it would then become a **'Boost'** charge. If the battery is allowed to be flat for long lengths of time, plates crystalize (sulphation) which may become permanent. This trickle charge would be called a **Constant Current** and **Constant Voltage Charge (CCV)** which could be around 15 hours from flat for a standard battery.

Boost charging for long periods can harm the battery as the plates could get hot. Wet batteries have a finite life of several years after which the plates slowly lose their Lead Material which sinks to the bottom of the battery as sediment. This action can be hastened with things like bad maintenance or vibration. In a glass battery you will see this. Of course the more sediment then the more inefficient the battery. Remember that if the lead sediment builds up in the bottom of the battery it can, under extreme circumstances, short out the plates.

Reverse Polarity:

There is a phenomenon on a secondary battery which can happen even though extremely unlikely and that is reversing the polarity of the battery. If you connect the battery charger the wrong way round on a typical Lead Acid battery this is unlikely to reverse the polarity. You may not even see a spark, unless on boost, from the charger leads although you may see the voltmeter read even when the charger is switched off as you connect it. This may damage the battery and it would not charge.

When jump starting a car from another battery it is very different. You should connect up one lead and quickly just touch the other lead on the terminal without clipping it on. There will be a huge spark if you have reverse polarity and one battery could explode if you clipped the clamp onto the terminal.

The only way that I can think of, in theory, where you would actually reverse the polarity of the battery is if you came across a completely flat battery and connected the charger the wrong way round and charged from completely flat.

Topping Up the Electrolyte:

Maintenance would be checking the specific gravity (SG), topping up with distilled or deionised water **(NOT TAP OR BOTTLED WATER)** and cleaning any powder off the electrodes. When checking the SG do not check when the battery has just been topped up with water as the water will sit on top of the acid for a while.

Electrolyte liquid should never be allowed to drop below the plates otherwise, again, they will sulphate over and become inefficient and may become too permanent to reverse, but on the other hand do not overfill or the SG will be wrong and the electrolyte mixture will be too thin. Never add acid: this should not change from manufacture. Only top the battery up to cover the plates **BEFORE** charging and finish filling correctly **AFTER** charging.

Gel Electrolyte:

It is possible to obtain 'Gel' Lead Acid batteries where the electrolyte is gel instead of liquid. This will cut maintenance down but not stop it completely. This type of battery, not being liquid, is mostly spill proof. Currents tend to be higher when charging and discharging.

Sub Divisions of 'Types' of Battery:

These may sometimes come in the form of an **Enhanced Flooded Battery (EFB)** or **Absorbent Glass Mat (AGM)** which differ from each other slightly in the way of performance. **Deep Cycle Batteries** differ again slightly from the battery used in a car although the chemistry may be similar. Car batteries give large amounts of energy in one burst where as **Deep Cycle** batteries give their energy over time which is ideal for solar energy! Deep cycle batteries would have a higher Amp hour (Ah) rating than a car battery.

Stratification:

This is where the concentrated Sulphuric Acid moves to the bottom of the battery and the watery part of the electrolyte moves to the top. All to do with specific gravity and could always be a problem for the time being with wet/flooded batteries. Usually the water and acid mix under motion conditions and a phenomenon called 'gassing' which is usually achieved from the car's alternator charging system. If, under very extreme conditions, the natural gassing does not work then manual gassing is used under controlled conditions.

Sealed Batteries/Maintenance Free:

Sealed or maintenance free batteries are exactly what they say and electrolyte level and specific gravity checks cannot be carried out on these batteries. Voltage checks under load can be done.

Lead Acid sealed/maintenance free batteries are usually guided by weight. The lighter the battery the thinner the plates and less Lead. The heavier the battery the more Lead on the plates. So you may find the lighter battery is cheaper because of less Lead, but will not last as long as the heavier one with more Lead. It depends what is wanted. Manufacturer's data gives battery life.

Additives:

It is possible to obtain additives to add to battery electrolyte to reduce sulphation and improve the internal condition of the battery or reduce the internal resistance. Two of these additives are Epsom Salts and Tetrasodium Ethelenediamintetraacetate (EDTA). Epsom Salts & EDTA are supposed to dissolve sulphation. Whether or not they work I have no idea although it may be a last resort!

BE VERY CAREFUL BREATHING IN FUMES FROM ETHELENEDIAMINTETRAACETIC ACID. IRRITATES SKIN, NOSE, THROAT & EYES. ALSO MAY DAMAGE KIDNEYS. READ LABELS CAREFULLY!

Secondary Lead Acid Battery DO NOTS:

1 – DO NOT add acid. Acid does not evaporate, although acid mist can be present when caps are removed, but the water will evaporate, sometimes due to high temperature or overcharging. Just for interest, water boiling point 100 degrees centigrade, acid boiling point 340 degrees centigrade. Only top up with distilled or deionised water, **NOT TAP OR BOTTLED WATER!**

2 – DO NOT add too much water over the recommended level. Remember that too much water will dilute the electrolyte too much and will change the specific gravity (SG) and the battery will become inefficient. The electrolyte could overflow when charging due to heat.

3 – DO NOT let the liquid level go below the plates otherwise after long periods of exposure the sections of the plates will sulphate and that part of the plate will become inoperative cutting down the efficiency of the battery.

4 – DO NOT add Sulphuric Acid. If the SG exceeds the recommended value, corrosion inside of the battery may increase. Check with manufacturer.

5 – DO NOT use equipment for **alkaline** batteries on **acid** batteries and visa-versa. Any alkaline contact with acid will cause a neutralisation effect and cut down the efficiency.

6 – DO NOT allow the battery to discharge too much for long periods of time otherwise 'sulphation' (Lead Sulphate crystal deposits on the plates) takes place and the life of the battery will be shorter than if a charge was applied every so often to keep the voltage up. If the battery charging system is efficient the Lead Sulphate crystals are regularly converted back.

7 – DO NOT overcharge the Lead Acid battery. This is easily done if the battery is low and you put it onto trickle charge and forget about it. Unlike float charge, trickle charge will just keep charging at the same current and eventually the overcharging can cause the Lead plates to heat up and swell due to Lead having a high expansion rate.

If the plates swell the electrode compound can fall off into the bottom of the battery and in extreme circumstances cause short circuits inside of the battery and explosions or shorter life span.

8 – DO NOT ignore batteries that are stored. At room temperature (around 15-20 degrees centigrade) the battery 'self' discharge' could be up to 5% per month and **SHOULD NOT** be left for more than 6 months without attention. Always refer to manufacturer's instructions on storage of unused batteries.

9 – DO NOT short circuit the battery to check if it is charged. It has a very high short circuit current. Use the correct voltage tester with resistances (battery = 12 volts, shorting wire = 0.4Ω – using Ohm's Law V/R = I so 12/0.4 = **30Amps** or quite a bang).

10 – DO NOT try to remove corrosion on a Lead Acid battery without proper PPE meaning at least eye protection and rubber type gloves. This corrosion will usually be on lead acid battery anodes that are not sealed.

The most common corrosion inside the battery is called sulphation (as earlier) and outside the white powder on the anode terminal can be Lead Sulphate or Copper Sulphate & Acid, which is usually a reaction between the sulphate in the battery and the Lead post/Copper cable connection. Like the powder on the small cylinder batteries it is highly corrosive and can be a hazard to eyes and skin.

Usually the battery with this corrosion is of the older type, but it can be caused by overcharging or undercharging and the battery releasing gas. Baking soda should neutralise the acidity of the white powder so scrub with a small brush, but do not douse the top of the battery with tap water, it could cause a short circuit. You may actually think that you could use distilled water to wash the top being an insulator, but once dirt/impurities mix with the distilled water it may become a conductor and hence short circuits.

AGAIN I MUST STRESS HERE: USE EYE PROTECTION & RUBBER GLOVES AT LEAST!

Secondary Nickel Batteries (General):

Primary & Secondary Batteries: As we discussed earlier in the book, primary batteries have a distinct life span so when they are flat that is the end of their life and they are then disposed of, but secondary batteries can be recharged from flat to nearly what they were originally. Nickel type batteries are mainly secondary batteries so chemical reactions and ion flow can be reversed.

Battery Sections: As with every other battery these can be divided into sections i.e.: a container which in this case is can be plastic, steel etc., an anode (negative) electrode and cathode (positive) electrode which contain for instance Nickel and other elements, an electrolyte of usually Potassium Hydroxide (KOH) and sometimes containing what is known as a separator.

Separator: The separator is there quietly sitting in the electrolyte mainly to keep the anode from touching the cathode and causing a short circuit inside of the battery. The separator must be porous enough to still allow any chemical reactions or ion flow to take place inside the battery.

Anodes & Cathodes: Anodes in Nickel batteries can be metal i.e.: Iron, Zinc or Gas such as Hydrogen or Nickel Oxide. Cathodes can be Nickel Oxide in many cases, or Nickel Hydroxide, or in one case pure Nickel.

Lifespan: Usually Nickel batteries, as with other rechargeable batteries, have a finite life of a few years. Shelf life and use should be taken into consideration because both of these values has an effect on the life of the battery. When a rechargeable battery is charged it sometimes never returns to the same charge that it had when it left the manufacturers.

Electrolyte: Most of the Nickel batteries that we are going to mention have Potassium Hydroxide (KOH) as the electrolyte. We all have electrolytes inside of our bodies controlled by diet to conduct impulses from the nervous system. This may be called our **pH Level**.

Electro Chemical Potential: The ability of a metallic material anode to lose electrons. Lithium of all the elements has a much higher tendency to lose electrons than Nickel.

Thermal Runaway: A thermal runaway can be termed as an **'Uncontrollable Exothermic Reaction'**. Let us say that the battery becomes internally damaged to, say, short circuit stage. The amount of heat generated inside the battery would be enormous and unstoppable. This is why the above separator is in place in many cases.

Battery Container: The battery container can be made of Zinc, Steel, Plastic etc. depending upon the type of Nickel battery and electrolyte.

Nominal Voltage: This is the voltage of the battery when it is half way between being fully charged and very low charge.

Open Circuit Voltage: This is the voltage of the battery with no load. So if you picked up a new battery and put a multi-meter onto it then this is the voltage it would read. Standard alkaline batteries AA, AAA, C and D, are **ALL** around 1.5 volts each. The different Nickel batteries, however, may produce between **1.2–1.7 volts.**

Uses: Have been used in older electric vehicles. Many are the standard rechargeable cylinder batteries that you buy every day in the form of AA, AAA, C and D batteries. Larger Nickel batteries are sometimes used in **uninterruptable power supplies (UPS).**

Energy Density: How small and what weight the battery is in mass compared to its steady power output. Usually worked out in **Watt hours/Kilogram (Wh/Kg).**

Life Cycle: Even rechargeable batteries have a finite life span which is called the **Life Cycle.** The life cycle is the amount of charges/discharges the battery can stand before its efficiency drops below 80% and the battery will never after reach full charge. The life cycle depends upon the type of Nickel battery you have. This could be, for example, 2000 times over a number of years, say, 3 years.

Amp hours: Battery Amp hours (Ah) are found by multiplying how many amps the battery is capable of supplying **(Amps)** by the discharge time **(the time it takes the battery to go flat at maximum output in hours).** So if a typical Nickel battery output can supply a current of 10 amps for 20 hours then it is a 200Ah (Amp hour) battery (10 x 20 = 200).

Power Density: Very slightly different from **Energy Density.** Energy Density deals with a steady output compared to its mass, whereas **Power Density** is how much power can be extended in one period compared to its size and mass. So for example, a battery with a high energy density can have a low power density and supply an appliance for longer periods. Back to our electric car!

Battery Discharge: Nickel-based batteries are a typical chemical reaction. There are certain elements in the electrodes which, when the battery is working, i.e.: there is a load connected to it, slowly react with the electrolyte to form a different set of chemicals. One analogy might be you obtaining a container and adding three liquid chemicals, representing each of two electrodes and an electrolyte, and mix them all together by stirring. To start with you have three definite substances, but by stirring you form another, different, chemical. Well this is very similar to a battery discharge except the 'stirring' is natural and much slower and produces DC electricity on the way. When all the chemicals are mixed, the battery is flat.

Battery Charge: Let us take the above analogy. We have a container full of mixed chemicals and now we have to change them back to what they were to start with and this is where the charging comes in. Many Nickel batteries of the rechargeable type must have a safety valve mechanism at the top to allow gases to escape which might be made during charging or overheating. The charger must be of battery manufacturer type and not home made in any way. I must state here that the gases may **USUALLY** be released in large quantities on overcharging rather than normal charging. As the battery ages the % of manufacturer's charge will drop.

Speed of Charge: With electric vehicles one of the most important points challenging the manufacturers is **'Speed of Charge'**, i.e.: how long does it take to get the battery back up to 100% charge. Let us take a petrol car. You run low on fuel and pull into a petrol station and fill up. How long does it take you? 10 minutes? If you pull into a station with a battery powered vehicle how long does it take to fill up with charge? Hours? Not acceptable so this has to be faster as with your mobile phone, iPad etc.

Over Charge: Extra energy produced in overcharge sometimes manifests itself in the form of heat. Can I overcharge a Nickel battery? The short answer is **YES** so try to avoid boost charging and stick to trickle charge. Certain cells have internal built in safety devices which stop internal gases generated by charging from doing any harm. Some of these safety devices are based on pressure, others on temperature. It is always wise to check with the manufacturer's data.

Over Discharge: Can I overcharge Nickel batteries? The answer is **YES**. **Nickel Metal Hydride** batteries, for instance, may self-discharge over a period of time so top up charges may be required. The cell voltage may drop from 1.2 volts to 1 volt below which the battery may suffer serious damage. If the damage does occur you may find that the cell will not charge to full capacity, if at all.

Specific Gravity: SG is measured with a **Hydrometer** and is a measure of the battery's charge at that moment in time. What we are measuring in actual fact is the electrolyte. A battery starts off with three basic chemicals i.e.: an **Anode,** a **Cathode** and an **Electrolyte**.

As the battery discharges the electrolyte starts to change from one medium to another until the battery is flat and the new chemical change will have a different SG. We then charge the battery and change the present medium back to the original three chemicals. What we are measuring is the weight of the electrolyte as it changes to another form. So how far on with that change you are will determine the charge of the battery.

Memory: The **Nickel Cadmium Batteries** especially are known for their **'Memory' Effect** so if the battery charge is regularly only partially used before charging it will remember and not give any more. Let us say that you use 35% of the battery total charge and then recharge to 100%. In time the battery will remember and will not give you beyond 35% of its total charge. Latest batteries do not have this 'memory' effect and of course this does not apply to the lithium battery.

Operating Temperature: Nickel batteries have a temperature in which they operate which will be determined by the manufacturers having carried out numerous tests. If the user temperature was outside these parameters then the battery life, charge and efficiency would be affected.

Safety Positive Temperature Co-efficient (PTC): One safety device that many secondary batteries have is what is called a **Positive Temperature Co-efficient (PTC)**. This is a type of resistance switch that works by temperature. The resistance is low under normal conditions, but if the temperature was to rise significantly, this would go high resistance and stop the battery working. It would return to normal as the battery cooled down.

Safety Gas Venting System: Some batteries have this system. It is there to allow gases inside of the battery produced during charging to vent out. The actual container is usually very strong so that these gases do not cause it to 'bulge' or rupture, but in extreme cases batteries have been known to explode even with a strong case.

Safety Current Interrupt Device (CID): This is a weak point in the battery, like a fuse, that will melt if the battery current gets too high. This is a safety device to stop the battery exploding due to excess current. One of the human failings that could cause excess current is to mix battery types. If one battery was to go flat quicker than the other(s) it could cause the other batteries to produce excess current in their effort to keep the appliance working.

Battery Management System (BMS): This is usually a system within a very sophisticated charging system that ensures that the battery cannot overcharge and will endeavour to keep the charging current at a constant flow.

Jelly Roll: This is the way the manufacturers make the **Cylinder Battery**. They put all the ingredients together along with separators, electrodes and electrolyte in a flat layered form and then role it into a tube like seaside rock and this action is called a **'Jelly Roll'**. The objective is to ensure that the electrodes are more in contact with the electrolyte.

Storage: Must be done to manufacturer's instructions. For instance the Nickel Iron battery must be stored flat. The storage temperature for **Nickel Metal Hydride Batteries** is between -20 to +35 degrees centigrade.

If we take **Lithium Batteries**, they should be stored at around 10 degrees centigrade. Storing at different temperatures to those recommended will be detrimental to the life of the battery. Some secondary batteries will self-discharge during storage and may require occasional charge. The battery should be stored without dropping more than around 50% of its charge.

Disposal: Batteries contain many chemicals and elements that, as explained earlier, we must prevent at all costs from getting into the food chain and water table through things like landfill.

There are UK Regulations in place for the disposal of batteries to protect the environment by not having them end up in a landfill. The regulations cover 3 types of battery, namely: 1 – the common car battery, in every car, that works the general electrics, 2 – the battery that powers electric vehicles, 3 – the batteries that power torches etc.

The regulations are: Waste Batteries and Accumulator Regulations 2009 and the UK department that enforces the regulations is The Office for Product Safety & Standards (OPSS).

Primary & Secondary Nickel Batteries:

There are probably more Nickel batteries on the market than people think, powering our electric cars, mobile phones, laptops etc. This section gives a basic, general overview of the Nickel battery and how it might work.

Typical Nickel Battery (Left) & Nickel Hydrogen Fuel Cell (Right)

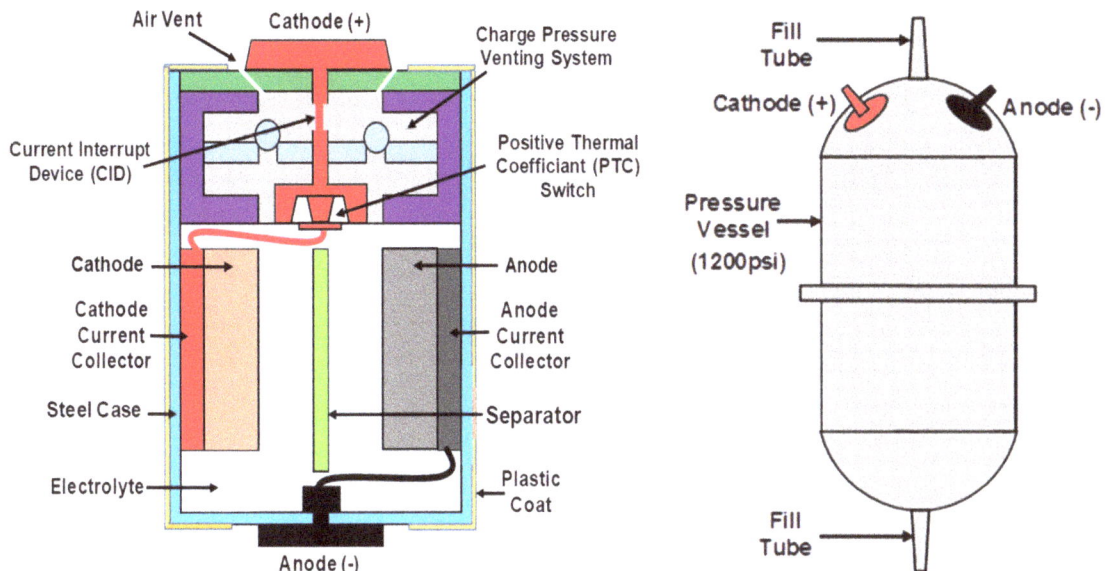

The diagrams above are of a typical Nickel battery (left) and the experimental Nickel Hydrogen Fuel Cell (right). I have drawn the Nickel Battery as a cylinder battery for ease of understanding. Not all Nickel batteries will have the safety devices in them that I have shown above. Fuel Cells differ from batteries because they have to take in a gas such as Hydrogen from the outside and expel bi-products whereas a battery is a self-contained unit of a case with a chemical reaction going on inside.

The Nickel battery has come on leaps and bounds in the electric car industry and is certainly on a par with Lithium batteries which are explained later in the book.

Let us have a look at the different types of Nickel batteries i.e.: what anode and cathode materials are used and what different electrolytes are in them.

1) Primary Nickel Oxyhydroxide Battery (See Primary Batteries)

2) Secondary Nickel Iron Battery (NiFe) (Edison Battery)

3) Secondary Nickel Metal Hydride Battery (NiMH) (MEMORY)

4) Secondary Nickel Zinc Battery (NiZn)

5) Secondary Nickel Cadmium Battery (NiCd) (MEMORY)

6) Secondary Nickel Hydrogen (NiH) Battery (Fuel Cell)

7) Secondary Nickel Cobalt (NiCo) Battery (Under Development)

8) Secondary Lithium-ion Nickel Cobalt Aluminium Oxide Battery (See Lithium)

9) Secondary Lithium-ion Nickel Manganese Cobalt Oxide Battery (See Lithium)

10) Secondary Nickel Zinc Single Flow Battery (See Flow Batteries)

Secondary Nickel Iron Battery (NiFe):

We can thank Thomas Edison for this battery as it is a direct descendant from his initial design.

Anode: Iron (Fe).

Cathode: Nickel Oxide – Hydroxide (NiO(OH)$_2$).

Electrolyte: Alkaline electrolyte is a mixture of Potassium Hydroxide (KOH) and distilled water (H_2O). There is also a porous separator included in the electrolyte which as we have mentioned keeps the anode from touching the cathode causing a short circuit.

Enclosure: The enclosure for the electrolyte and electrodes etc. in this case is made of inert Steel.

Voltage: The output voltage of each individual battery is **1.2 volts** with an open circuit voltage of **1.4 volts** so several batteries must be connected in a series/parallel connection to obtain the voltage required.

Battery Form: Wet/Flooded batteries that have caps and breathers and the electrolyte must be topped up with distilled or deionised water.

Weight: Batteries tend to be lighter than the lead acid battery.

Charge Temperature: -40 to +40 degrees Centigrade.

Charge: Hydrogen given off so room must be well ventilated and lighting certified. Will usually stand being over and under charged without damage.

Storage: Batteries must be flat for storage.

Self-Discharge: There will be a degree of around 10–20%/month self-discharge which is quite high really.

Environment: Environmentally friendly.

Battery life: 20–30years+

Secondary Nickel Metal Hydride Battery (NiMH) (MEMORY):

The nearest battery design is Nickel Zinc battery.

Anode: Made of a material that absorbes and stores Hydrogen (H) i.e.: Metal Hydride (MH)

Cathode: Nickel Oxide Hydroxide (NiOOH)

Electrolyte: Alkaline Aqueous Potassium Hydroxide (KOH) with a small amount of added Lithium Hydroxide. There is also a porous separator included in the electrolyte which as we have mentioned keeps the anode from touching the cathode causing a short circuit.

Power: Have many times the capacity of NiCd batteries.

Voltage: There is a lower voltage here than a conventional battery at **1.2 volts**.

Battery Form: Can be obtained in cylinder form such as AA, AAA, PP3 etc. Larger wet/flooded batteries also used in the early stages of hybrid electric vehicles, but now of course Lithium and Nickel batteries are the flavour of the day in electric vehicles because of their efficacy.

Environment: Environmentally friendly.

Bi-polar Models: Meaning where there is a solid electrolyte being developed. These may be called **'Solid State'** batteries.

Full Discharge: Can lead to cells going into reverse polarity which of course would be catastrophic. Can be subject to self-discharge even with no load so must be regularly monitored.

Nickel Cadmium Battery (NiCd) (MEMORY)

Anode: Metallic Cadmium (Cd) with a Cadmium Hydroxide (Cd(OH)$_2$) Mix.

Cathode: Nickel Hydroxide (Ni(OH$_2$) or Nickel Oxyhydroxide (NiOOH) Mix.

Electrolyte: Alkaline, as the Nickel Metal Hydroxide battery, Aqueous Potassium Hydroxide (KOH).

Voltage: Again low at 1.2 volts.

Memory Effect: The older NiCd batteries are known for their **'Memory'** effect so if the battery charge is regularly only partially used before charging it will remember and not give any more. Let us say that you use 35% of the battery total charge and then recharge to 100% regularly; in time the battery will remember and will not give you beyond 35% of its total charge.

This effect is caused by Cadmium crystals forming inside the battery. So to sum up the battery must be flat before it is recharged (by flat I mean down to 1 volt/cell) so if you have an older device with one of these batteries powering it then let the device work until the screen goes off and then charge it.

Latest batteries do not have this 'memory' effect and of course this does not apply to the Lithium Battery.

Environment: These batteries are not too environmentally friendly because of Cadmium.

Battery Form: Can also be obtained in cylinder form such as AA and AAA etc. and the larger batteries are used in aviation.

Specific Gravity: Around 1.5.

The chemical formula for this NiCd Secondary Battery is:

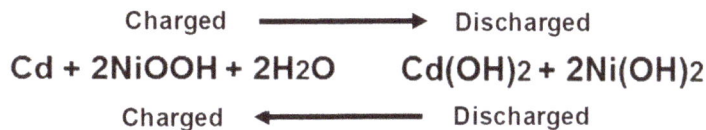

$$\text{Charged} \longrightarrow \text{Discharged}$$
$$Cd + 2NiOOH + 2H_2O \quad Cd(OH)_2 + 2Ni(OH)_2$$
$$\text{Charged} \longleftarrow \text{Discharged}$$

Nickel Hydrogen (NiH) Battery:

Getting towards being a **Fuel Cell** rather than a battery.

Anode: Made up of gaseous Hydrogen (H) at 1000+ psi.

Cathode: Nickel (Ni).

Electrolyte: Alkaline Potassium Hydroxide (KOH) like many Nickel based batteries.

Voltage: The voltage of this battery/fuel cell is a little over 1.5 volts.

Uses: Many highly sophisticated pieces of equipment such as satellites. The fuel cell of the future may look similar to the diagram at the beginning of this section rather than a conventional battery.

Self Discharge: High.

Nickel Cobalt (NiCo) Battery:

This battery is **under development** at the moment and seems to be one of the batteries of the future.

Anode: Cobalt Hydroxide (Co(OH)$_2$).

Cathode: Nickel Oxide (NiO).

Voltage: The voltage of this battery is 1.2 volts.

Environment: Non Flammable and environmentally friendly, but **Cobalt** mining might be questioned.

Secondary Nickel Zinc Battery (NiZn):

The nearest battery design is Nickel Metal Hydride battery.

Anode: Zinc (Zn).

Cathode: Nickel Oxide - Hydroxide ($NiO(OH)_2$).

Electrolyte: Alkaline Potassium Hydroxide (KOH). There is also a porous polymer separator included in the electrolyte which, as we have mentioned, keeps the anode from touching the cathode causing a short circuit. It also to a certain extent prevents dendrite.

Voltage: There is a higher voltage here at **1.7 volts** per cell and so this will cut down the number of cells required. They do not particularly like trickle charge. Cell will hold its charge until the last minute, but will self-discharge when not in use.

Battery Life: Shorter than other Nickel based batteries.

Dendrites: Sometimes a problem as they are small crystal whiskers that grow on the anode electrodes of metal batteries increasing the surface area.

Battery Form: Can be obtained in cylinder form such as AA and AAA and also larger wet/flooded form, sealed or top up. The larger batteries have caps and breathers and the electrolyte must be topped up with distilled or deionised water.

Environment: Non-flammable and environmentally friendly.

Secondary Lithium-ion Nickel Cobalt Aluminium Oxide Battery (NCA):

NCA: Nickel Cobalt Aluminium ($LiNi_{0.8}Co_{0.2}O_2$).

Anode: Carbon(C)/Graphite(C).

Cathode: Lithium Nickel Cobalt Aluminium ($LiNi_{0.8}Co_{0.2}O_2$).

In this battery there is not as much Cobalt as a NMC battery (below) so it would be lighter. Anodes and cathodes of Lithium based batteries usually have current collectors attached which would be Copper in this case on the anode and Aluminium on the cathode.

Electrolyte: Gel with a Lithium Salt added in the form of Lithium Hexafluorophosphate ($LiPF_6$).

Voltage: The nominal voltage is around **3.7** volts.

Secondary Lithium-ion Nickel Manganese Cobalt Oxide Battery (NMC):

NMC: Nickel Manganese Cobalt Oxide ($LiNiMnCoO_2$).

Anode: Carbon (C).

Cathode: Lithium Nickel Manganese Cobalt Oxide ($LiNiMnCoO2$) or Nickel Cobalt Lithium Aluminate ($LiNi_xCo_yAl_zO2$) (NCA). More Cobalt than the NCA battery.

Anodes and cathodes of Lithium based batteries usually have current collectors attached which would be Copper in this case on the anode and Aluminium on the cathode.

Electrolyte: Liquid or gel with a Lithium Salt added.

Voltage: The nominal voltage is **3.6–3.7 volts.**

Battery Life: Short.

Secondary Nickel Zinc Single Flow Battery: See Flow Batteries

Secondary Lithium-ion Batteries (General):

Primary & Secondary Batteries: As we discussed earlier in the book, primary batteries have a distinct life span so when they are flat that is the end of their life and they are then disposed of, but secondary batteries can be recharged from flat to nearly what they were originally. Lithium type batteries are mainly secondary ion batteries so chemical reactions and ion flow can be reversed.

Battery Sections: As with every other battery these can be divided into sections i.e.: a container which in this case is usually steel, an anode (negative) electrode and cathode (positive) electrode which contain for instance Lithium and other elements, and an electrolyte which can be liquid, gel or solid, sometimes containing what is known as a separator.

Separator: The separator is there quietly sitting in the electrolyte mainly to keep the anode from touching the cathode and causing a short circuit inside of the battery. The separator must be porous enough to still allow ion flow and chemical reactions to take place inside of the battery. Remember some of these Lithium batteries may be quite thin and flexible. Solid electrolytes do not obviously require a separator.

What are ions?: Particles called **ions,** some of which are called cations which, unlike natural atoms, have more protons than electrons so they are positive and other **ions** called anions which have more electrons than protons so they are negative, gather on each electrode.

Lithium Anodes & Cathodes: Lithium is a very soft metallic, alkali, chemical element with chemical symbol (Li) and is a very good conductor of electricity and may be used in the cathode whereas anodes in a lot of cases are carbon or graphite as the ions can easily be stored in the layers that make up the material. They first started Lithium battery development in the 1970s. There may actually be more types than you think.

Carbon/Graphite Anodes & Cathodes: Obviously Carbon (C) is a good conductor of electricity and can hold a lot of energy and has the chemical symbol (C). Graphite, from the Greek 'graphein' is a carbon allotrope which means that its composition is similar to carbon although it might look slightly different. Graphite also has the chemical symbol (C). The new material is **'Graphene'**, another allotrope of Carbon, but with many more enhanced properties so watch this space!

Anode & Cathode Electrodes: these are connected to the anode and cathode material and may be called '**Current Collectors**' the anode will usually be Copper and the cathode Aluminium.

Lifespan: Usually Lithium, as with other rechargeable batteries, have a finite life of a few years. Shelf life and use should be taken into consideration because both of these values have an effect on the life of the battery. When a rechargeable battery is charged it sometimes never goes back to the same charge that it had when it left the manufacturers.

Electrolyte: Some of the Lithium batteries that we are going to mention have solid electrolyte, some have liquid electrolyte and some have gel electrolyte. This medium is there to allow the flow of only charged ions to go from the anode to the cathode. We all have electrolytes inside of us controlled by diet to conduct impulses from the nervous system. This may be called our pH Level.

Electro Chemical Potential: The ability of a metallic material anode to lose electrons. Lithium of all the elements has the highest tendency to lose electrons.

Thermal Runaway: A thermal runaway can be termed as an '**Uncontrollable Exothermic Reaction'**. Let us say that the battery becomes internally damaged to, say, short circuit stage; the amount of heat generated inside of the battery would be enormous and unstoppable. This is why the above separator is in place.

Battery Container: What is the container made out of that holds the electrodes, electrolyte etc.? In some cases this container is made out of steel with a plastic coat.

Nominal Voltage: This is the voltage of the battery when it is half way between being fully charged and very low charge.

Open Circuit Voltage: This is the voltage of the battery with no load. So if you picked up a new battery and put a multi-meter onto it then this is the voltage it would read. Standard Alkaline batteries, AA, AAA, C and D, are **ALL** around **1.5 volts** each. The different Lithium ion batteries, however, may produce **1.5–3.7 volts.** So advice would have to be obtained as to where they can be used. They definitely are not suitable for every application.

Uses: These batteries are used in modern laptop PCs, smartphones, electric cars, power tools etc. The beauty of the Lithium ion battery is that it can remain very small and still give out enough energy to do whatever it is designed for. Bought individually, they can be very expensive. Manufacturers have to ensure that the batteries are safe at all times, let us say, in an electric vehicle in a collision as well as being environmental friendly.

Energy Density: How small and what weight the battery is in mass compared to its steady power output. Usually worked out in **Watt hours/Kilogram (Wh/Kg)**. Lithium batteries are ideal in this area as the power output is large compared to the size of battery. Back to our electric car!

Life Cycle: Even rechargeable batteries have a finite life span which is called the life cycle. The life cycle is the amount of charges/discharges the battery can stand before its efficiency drops below 80% and the battery will never after reach full charge. The life cycle depends upon the type of lithium battery you have. This could range from 800–5000 times over a number of years, say, 3 years. Does your mobile phone or iPad battery not last as long as it did when it was new?

Amp Hours: Battery Amp hours (Ah) are found by multiplying how many amps the battery is capable of supplying **(Amps)** by the discharge time **(the time it takes the battery to go flat at maximum output in hours).** So if a typical Lithium battery output can supply a current of 10 amps for 20 hours then it is a 200Ah (Amp hour) battery (10 x 20 = 200).

Power Density: Very slightly different from **Energy Density**. Energy Density deals with a steady output compared to its mass where **Power Density** is how much power can be extended in one period compared to its size and mass. So for example a battery with a high energy density can have a low power density and supply an appliance for longer periods. Back to our electric car!

Battery Discharge: In most cases the battery anode holds many charged Lithium-ion particles which are transported by the electrolyte to the cathode passing through what is called the perforated separator which allows the very tiny ions to pass through, sometimes gel or solid, which controls the reaction. The electrons flow through the appliance. This is sometimes referred to as a **'Redox'** **(Reduction Oxidation Reaction)**. Once **ALL** of the ions pass from the anode through the electrolyte to reach the cathode the battery internal reaction is stable i.e.: the battery is flat. The ions by charging must now be coaxed back from the cathode to the anode to start the process all over again.

Battery Charge: On battery charging the Lithium ions move, via the electrolyte, to where they were when the battery was fully charged. Many of this type of battery must have a safety valve mechanism at the top to allow gases to escape which might be made during charging or overheating. There may be a PTC switch in case of internal rise in temperature and a fuse type mechanism in case of high rise in current. The charger must be of battery manufacturer type and not home made in any way. I must state here that the gases may be released more on overcharging rather than normal charging.

Speed of Charge: With electric vehicles one of the most important points challenging the manufacturers is **'Speed of Charge'** i.e.: how long does it take to get the battery back up to 100% charge. Let us take a petrol car – you run low on fuel and pull into a petrol station and fill up. How long does it take you? 10 minutes? If you pull into a station with a battery powered vehicle, how long does it take to fill up with charge? Hours? Not acceptable so this has to be faster as with your mobile phone, iPad etc.

Lithium & Water: Lithium is similar to Sodium as it is very volatile and can burst into flames if exposed to water. We can stabilise this a little by making it into a **Metallic Oxide.**

Lithium & Environment: Most Lithium-ion batteries are very environmentally friendly and are recyclable.

Over charge: Can I overcharge a Lithium battery? Lithium batteries are difficult to overcharge as usually the charging systems have built in safety mechanisms to stop this. So can I charge my laptop PC overnight? The answer might be yes **BUT** it is possible for the charging system to malfunction and the result could be the battery heating up resulting in fire or explosion. It is possible for the manufacturers to fit a **Current Interrupt Device (CID)** which cuts the power if the battery current was to rise beyond parameters. This is usually a one way action and is the end of the battery's life.

Over Discharge: Can I over discharge a Lithium battery? Almost certainly! If the Lithium battery drops below the manufacturers recommended voltage then the life of the battery can dramatically reduce. This might come into the open in an appliance that has been left unused for some length of time or you store the battery.

Memory: Older batteries such as Nickel Cadmium can develop what is called a **'Memory'** and this is a severe detriment to the lifespan of the battery. If the battery charge is regularly only partially used before charging it will remember and not give any more. Let us say that you use 35% of the battery total charge and then recharge to 100%. In time the battery will remember and will not give you beyond 35% of its total charge. Latest batteries do not have this 'memory' effect and of course this does not apply to the Lithium battery.

Operating Temperature: Lithium batteries have a temperature in which they operate which will be determined by the manufacturers having carried out numerous tests. If the user temperature was outside these parameters then the battery life and efficiency would be affected.

Safety Positive Temperature Co-efficient (PTC): One safety device that many secondary Lithium batteries have is what is called a **Positive Temperature Co-efficient (PTC)**. This is a type of resistance switch that works by temperature. The resistance is low under normal conditions, but if the temperature was to rise significantly this would go high resistance and stop the battery working. It would return to normal as the battery cooled down.

Safety Gas Venting System: This system is there to allow gases inside of the battery produced during charging to vent out. The actual container is usually very strong so that these gases do not cause it to 'bulge' or rupture, but in extreme cases batteries have been known to explode even with a strong case.

Safety Current Interrupt Device (CID): This is a weak point in the battery, like a fuse, that will melt if the battery current gets too high. This is a safety device to stop the battery exploding due to excess current. One of the human failings that could cause excess current is to mix battery types. If one battery was to go flat quicker than the other(s) it could cause the other batteries to produce excess current in their effort to keep the appliance working.

Battery Management System (BMS): This is usually a system within a very sophisticated charging system that ensures that the battery cannot overcharge and will endeavour to keep the charging current at a constant flow.

Storage: Lithium batteries should be stored at around 10 degrees centigrade. Storing at different temperatures to those recommended will be detrimental to the life of the battery. Lithium batteries will self-discharge during storage and may require occasional charge. The battery should be stored without dropping more than around 50% of its charge.

Secondary Lithium ion Batteries:

Typical Lithium ion Battery

The diagrams below are of a typical Lithium Battery. They show the **Ion Flow** during **discharge** i.e.: when the battery is supplying power and the **Ion Flow** when the battery is being **charged.**

Discharge   **Charge**

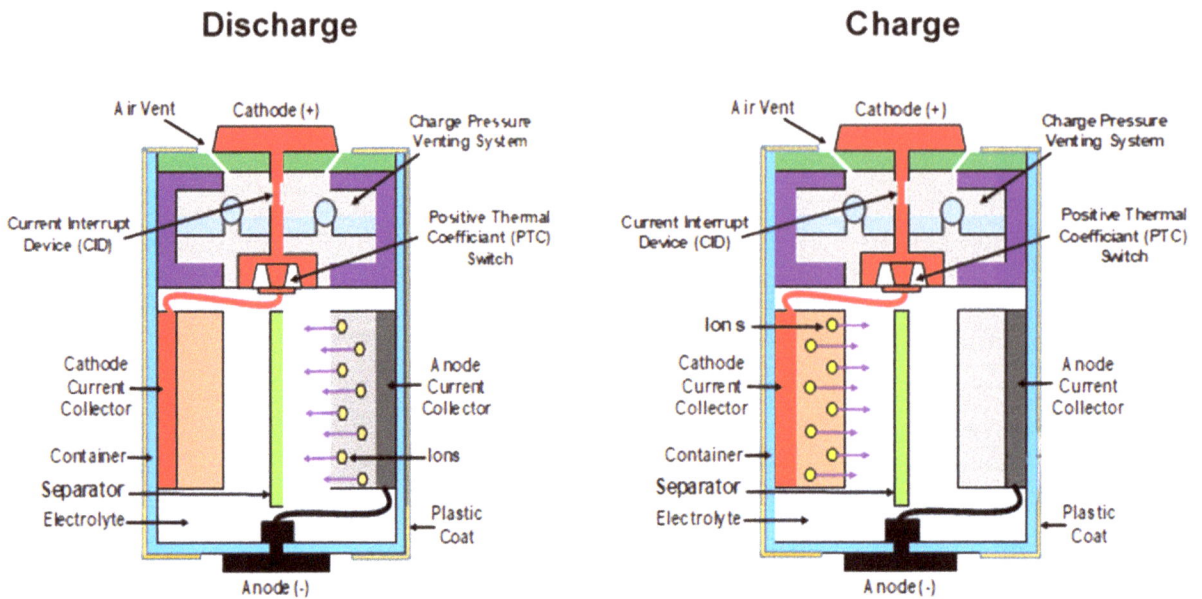

Types of Lithium Battery:

1) Lithium-ion Nickel Cobalt Aluminium Oxide Battery (NCA):

2) Lithium-ion Nickel Manganese Cobalt Oxide Battery (NMC):

3) Lithium-ion Cobalt Oxide Battery:

4) Lithium-ion Manganese Oxide Battery (LMO):

5) Lithium-ion Titanate Battery (LTO):

6) Lithium-ion Iron Phosphate Battery (LFP):

7) Lithium-ion Polymer (Flat Shape) Battery:

8) Lithium-ion Thin Film Battery:

9) Lithium-ion Air Battery:

10) Lithium-ion Sulphur Battery:

11) Lithium-ion Silicon Battery (Sand Battery):

12) Lithium-ion Ceramic Battery (FLCB or PLCB):

13) Lithium-ion Carbon Dioxide Battery:

14) Lithium-ion Hydride Battery:

There are descriptions of the various Lithium batteries on the following pages.

Secondary Lithium-ion Nickel Cobalt Aluminium Oxide Battery (NCA):

Anode: Carbon(C)/Graphite(C).

Cathode: Lithium Nickel Cobalt Aluminium (LiNi$_{0.8}$Co$_{0.2}$O$_2$).

In this battery there is not as much Cobalt as a NMC battery (below) so it would be lighter. Anodes and cathodes of Lithium based batteries usually have current collectors attached which would be Copper in this case on the anode and Aluminium on the cathode.

Electrolyte: Gel with a Lithium Salt added in the form of Lithium Hexafluorophosphate (LiPF$_6$).

Voltage: The nominal voltage is around **3.7** volts.

Secondary Lithium-ion Nickel Manganese Cobalt Oxide Battery (NMC):

Anode: Carbon (C).

Cathode: Lithium Nickel Manganese Cobalt Oxide (LiNiMnCoO2) or Nickel Cobalt Lithium Aluminate (LiNixCoyAlzO2) (NCA). More Cobalt than the NCA battery.

Anodes and cathodes of Lithium based batteries usually have current collectors attached which would be Copper in this case on the anode and Aluminium on the cathode.

Electrolyte: Liquid or gel with a Lithium Salt added.

Voltage: The nominal voltage is **3.6–3.7 volts.**

Battery Life: Short.

Secondary Lithium-ion Cobalt Oxide Battery:

Anode: Carbon (C)/Graphite (C).

Cathode: Lithium Cobalt Oxide (LiCoO2).

Anodes and cathodes of Lithium based batteries usually have current collectors attached which would be Copper in this case on the anode and Aluminium on the cathode.

Electrolyte: Gel with a Lithium Salt added.

Works by what is called Intercalation (Cathode ion storage) and Dis-intercalation between the anode and cathode.

Voltage: 3.6 volts.

Lifespan: Lower cycle life than other Lithium batteries.

Secondary Lithium-ion Manganese Oxide Battery (LMO):

Anode: Lithium Manganate Oxide (LiMn2O4).

Cathode: Carbon (C)/Graphite(C).

Anodes and cathodes of Lithium based batteries usually have current collectors attached which would be Copper in this case on the anode and Aluminium on the cathode.

Electrolyte: Lithium Salt added in the form of Lithium Hexafluorophosphate) (LiPF$_6$).

Voltage: The nominal voltage is **3.7 volts.**

Lifespan: Boost charging regularly will reduce lifespan dramatically as it may destroy the internal lattice.

Secondary Lithium-ion Titanate Battery (LTO):

Anode: Lithium Titanate coated with Nanocrystals giving it an enormous surface area. **Titanate** is an inorganic compound of **Titanium Oxide**.

Cathode: Lithium Manganese Oxide (NMC) or Lithium Manganate Oxide ($LiMn_2O_4$).

Anodes and cathodes of Lithium based batteries usually have current collectors attached which in this case is Aluminium on **both.**

Voltage: Lower nominal voltage than other Lithium batteries of around 2.4 volts.

Charge: Faster charge than some other Lithium batteries which makes them better for electric vehicles and UPS systems.

Lifespan: Good life cycle 20,000+.

Secondary Lithium-ion Iron Phosphate Battery (LFP):

Anode: Carbon (C)/Graphite (C).

Cathode: Lithium Iron Phosphate ($LiFePO_4$).

Anodes and cathodes of Lithium based batteries usually have current collectors attached which would be Copper in this case on the anode and Aluminium on the cathode.

Electrolyte: Liquid or gel of organic solvents with a Lithium Salt added.

Voltage: The nominal voltage of this battery is around **3.2 volts.**

Energy Density: Seems to have a lower energy density than other Lithium batteries of the same size.

Lifespan: No memory so partial use/charge will not affect the battery life.

Secondary Lithium Polymer (Flat Shape) Battery:

Anode: Lithium (Li).

Cathode: Lithium Cobalt Oxide ($LiCoO_2$).

Anodes and cathodes of lithium based batteries usually have current collectors attached which would be Copper in this case on the anode and Aluminium on the cathode.

Electrolyte: The electrolyte would be **Solid Polymer** in this case.

Voltage: The nominal voltage of this battery is around **3.6 volts.**

Secondary Lithium-ion Thin Film Battery:

Anode: Carbon (C)/Graphite (C).

Cathode: Lithium Cobalt Oxide ($LiCoO_2$).

Anodes and cathodes of Lithium based batteries usually have current collectors attached which would be Copper in this case on the anode and Aluminium on the cathode.

Electrolyte: Lithium Phosphorous Oxynitride (LiPON) layered so that it is flexible.

Voltage: The nominal voltage would be around **3.6 volt.**

Container: Container shell would be **Silicon Nitride (Si_2N_4).** Thin film batteries hence the name are so thin that they can be used for medical implants.

Secondary Lithium-ion Air Battery:

This is another variation of a Lithium-ion battery.

Anode: Lithium Metal (Li).

Cathode: Anode emits ions across an electrolyte and then, very similar to the hearing aid battery, air is allowed into the porous cathode and reacts with the positive charge ions to form **Lithium Peroxide (Li_2O_2)** which for better conductivity is separate from the positive electrode.

Anodes and cathodes of Lithium based batteries usually have current collectors attached which could be Copper on the anode and Aluminium on the cathode.

Separator: There must be a solid separator to stop the Lithium metal reacting too violently with the aqueous electrolyte.

Electrolyte: Earlier batteries were gel electrolyte where voltage and current were less. The electrolyte can be **Aprotic (Non Aqueous), Aqueous, or Solid.**

Voltage: Is around 3 volts.

Secondary Lithium-ion Sulphur Battery:

Lithium-ion Sulphur batteries, I believe, are the future. These batteries are under huge development at the moment. Sulphur is an abundant, cheap element. This battery is lighter than other batteries of the Lithium-ion types and the energy density is higher. The anode is **Metallic Lithium (Li)** with a **Sulphur (S) impregnated with Carbon (C)** cathode. These elements on the anode are under development at the moment and many elements are being tried. There is an organic electrolyte containing elemental Sulphur. The voltage is around 2 volts. Planned to be used in electric vehicles. **(Watch this space.)**

Secondary Lithium-ion Silicon Battery (Sand Battery):

This looks like another battery of the future and sometimes called a **'Sand' Battery**. This battery employs a Silicon based Lithium alloy with a **Lithium Oxide (Li_2O_4)** cathode.

Anodes and cathodes of Lithium based batteries usually have current collectors attached which could be Copper on the anode and Aluminium on the cathode.

Electrolyte: A solid electrolyte is used which at the moment they are finding unstable. Also during charge ions can exceed capacity 300–400% and split the anode. These problems will be ironed out in the near future. Hopefully to be used in electric vehicles because of its cheap production.

Secondary Lithium-ion Ceramic Battery (FLCB or PLCB):

Much research is going on with these batteries. There are apparently two types, the **Flexible Lithium Ceramic Battery (FLCB)** and the **Rigid Lithium Ceramic Battery (PLCB).** The electrolyte in this case is **Solid Ceramic Lithium Metal Oxides.** Fast charges do not raise the temperature of the battery which makes it very safe compared to some. FLCB batteries are only centimetres thick and very flexible. In fact this battery can almost be folded in half very easily and still work! Lithium ceramic batteries can stand most types of abuse, overcharging and flattening without any harm at all to the battery.

Secondary Lithium-ion Carbon Dioxide Battery:

Research is being carried out on this battery at present. The anode consists of positive charged ions of Lithium (Li) and the cathode of negatively charged ions of Carbon (C). During the process Carbon Dioxide (CO_2) is converted to **Lithium Carbonate (Li_2CO_3)** and **Carbon(C).** There are problems at the moment with Carbon clogging up the ion transfer process,

Secondary Lithium-ion Hydride Battery:

Cathode made of Vanadium Oxide (V_2O_5).

Secondary Flow or Redox Batteries:

Another name for flow batteries is **'Redox' (Reduction Oxidation) Flow Battery.** These batteries are different from other batteries as they have pumps which pump round different fluids at either side of a membrane and the result is electron flow across the membrane. These batteries can be huge.

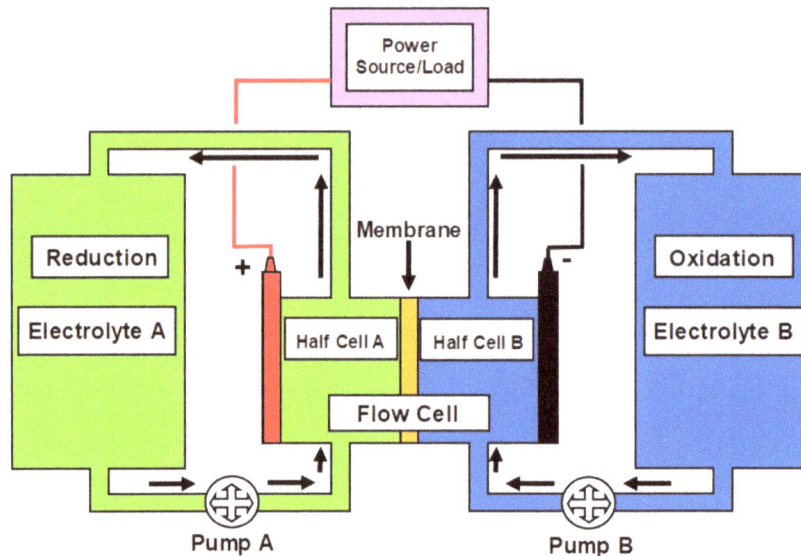

The two different electrolytes A (Green) and B (Blue), as can be seen in the diagrams above, are kept in their own containers and only enter the Cell/Battery half-cell sections when pumped round by pumps A and B and thereby cause an EMF through the load. They can be charged by power sources such as wind turbines or solar cells which charge up the electrons in the electrolyte tanks A (Green) and B (Blue).

The electrons then charge up in the blue electrolyte tank: this is called **'Oxidation'**. These charged electrons move into the green tank which is called **'Reduction'**. We have now got a **'Redox'** reaction which is the charge. If we want the reverse to happen and an EMF to flow we now turn on the pumps and pump the electrolyte through the Half Cells A and B of the Flow Cell. The electrons now flow back through the membrane.

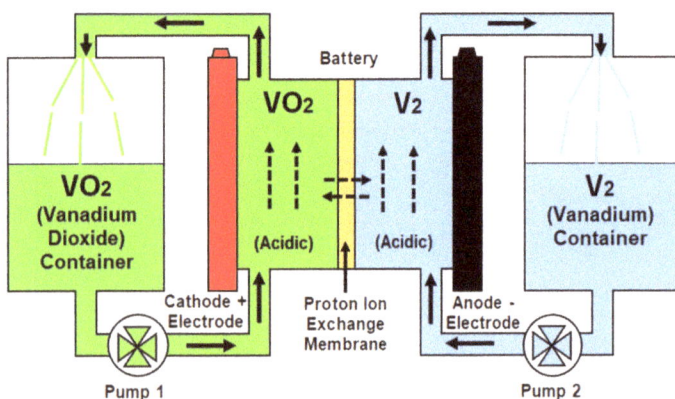

Electrolyte: In the diagram on the left there are two containers which hold the different electrolytes. The container on the left holds the **Vanadium Dioxide (VO2)** (Green) and the container on the right **Vanadium (V2)** (Blue). The electrolyte is acidic. **Pumps:** This battery has two pumps (1 & 2) and whilst the pumps are stopped the battery is dormant, apparently flat, and it will stay this way for considerable lengths of time.

Electrodes: These are red and black and are in contact with the fluid inside of two chambers (Green & Blue) which are fluid separated by a Proton ion exchange membrane.

Operation: To get this battery to give an output the pumps (1 & 2) must be started to achieve circulation of the two electrolytes and the produce is ion flow through the membrane and an output, voltage, from the two electrodes.

Voltage: Around 1.4 volts.

Zinc Bromine Flow Battery:

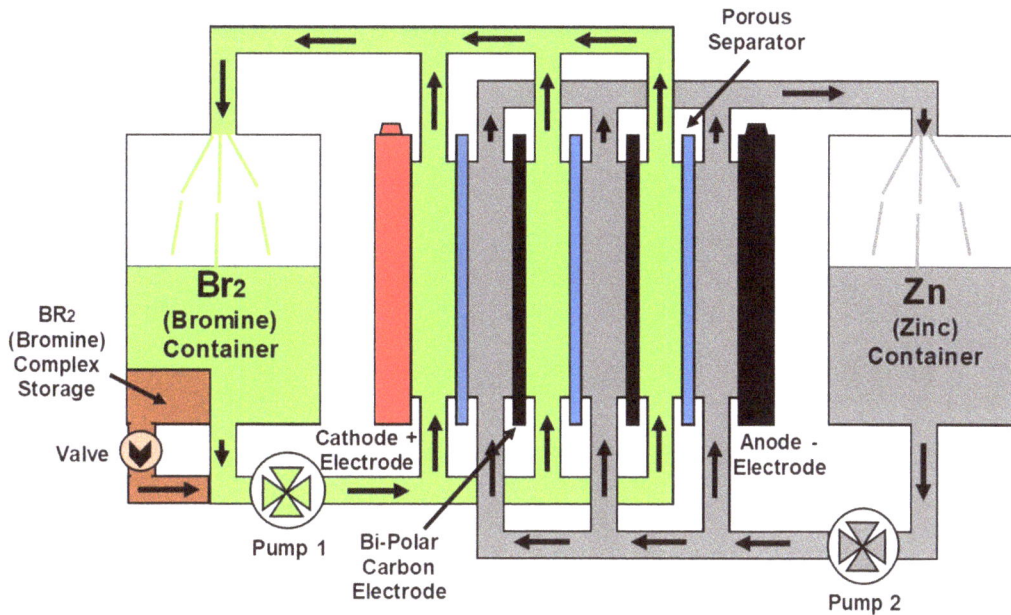

Electrolyte: In the diagram above there are two containers which hold the different electrolytes. The container on the left holds the **Bromine (Br2)** (Green) and the container on the right **Zinc (Zn)** (Grey).

Pumps: This battery has two pumps (1 & 2) and whilst the pumps are stopped the battery is dormant, apparently flat, and it will stay this way for considerable lengths of time.

Electrodes: These are red and black and are in contact with the fluid inside of the chambers (Green & Grey) which are fluid separated by porous separators and bi-polar carbon electrodes.

Operation: To get this battery to give an output, the pumps (1 & 2) must be started to achieve circulation of the two electrolytes and the produce is ion flow and an output, voltage, from the two electrodes.

Voltage: Around 1.8 volts.

Nickel Zinc Single Flow Battery:

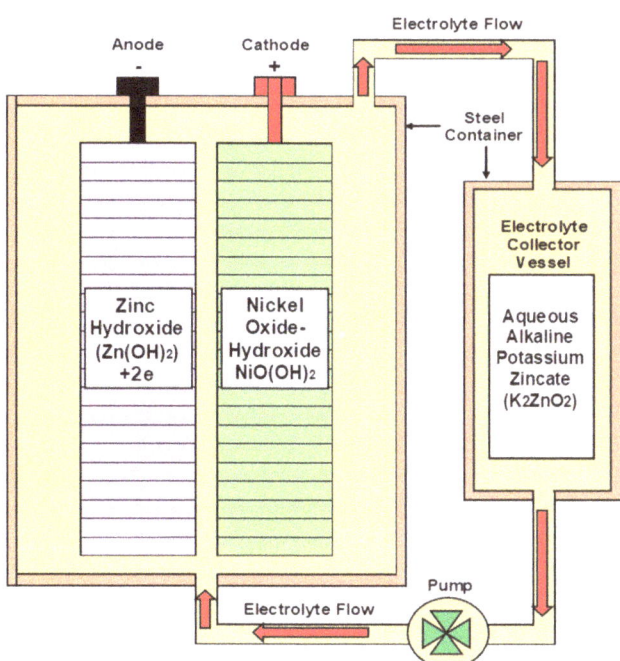

This Battery only requires one pump for the electrolyte and no separator required.

The positive electrode (Cathode) is **Nickel Oxide/Hydroxide NiO(OH)2.**

The negative electrode (Anode) can be **Zinc Hydroxide Zn(OH)2 +2e.**

The electrolyte is Aqueous Alkaline **(Potassium Zincate) K2ZnO2.**

Most flow batteries have two flow pumps, but this one only has a single pump, hence the name.

Polysulphide Bromide Flow Battery:

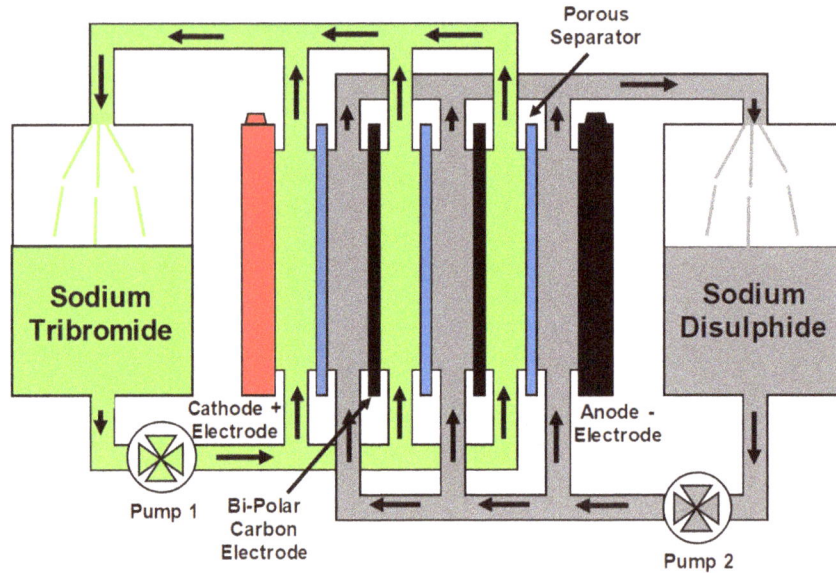

Electrolyte: In the diagram above it is very similar to the Zinc Bromine battery. There are two containers which hold the different electrolytes. The container on the left holds the **Sodium Tribromide (NaBr3)** (Green) and the container on the right **Sodium Disulphide (Na2S2)** (Grey).

Pumps: This battery has two pumps (1 & 2) and whilst the pumps are stopped the battery is dormant, apparently flat, and it will stay this way for considerable lengths of time.

Electrodes: These are red and black and are in contact with the fluid inside of the chambers (Green & Grey) which are separated by porous separators and bi-polar Carbon electrodes.

Operation: To get this battery to give an output, the pumps (1 & 2) must be started to achieve circulation of the two electrolytes and the produce is ion flow and an output, voltage, from the two electrodes black & red.

Zinc Cerium Flow Battery:

Charge: Redox as per model.

Electrolyte: In the diagram on the right there are two containers which hold the different electrolytes. The container on the left holds the **Cerium (Ce2)** (Brown) and the container on the right **Zinc (Zn2)** (Grey). The electrolyte is acidic.

Pumps: This battery has two pumps (1 & 2) and whilst the pumps are stopped the battery is dormant, apparently flat, and it will stay this way for considerable lengths of time.

Electrodes: These are red and black and are in contact with the fluid inside of two

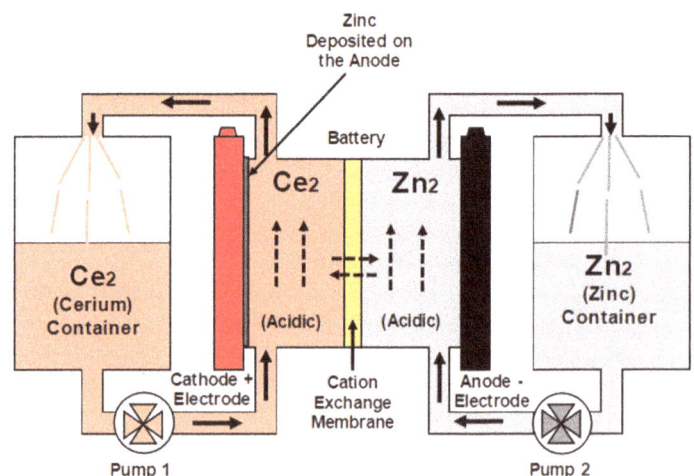

chambers (Brown & Grey) which are fluid separated by a **Cation Exchange Membrane.**

Operation: To get this battery to give an output the pumps (1 & 2) must be started to achieve circulation of the two electrolytes and the produce is ion flow through the membrane and an output, voltage, from the two electrodes.

Voltage: Around 2 volts.

Other Less Common Secondary Ion Batteries:

Lithium ion batteries at the moment are the flavour of the day. Many electric vehicles will be using this type of battery for its practicality and safety. Lithium ion batteries are not the only ion battery though and with much more research could be overtaken by other ion type batteries.

If you look at the history section you will see that the Lithium ion battery was invented and developed by John Goodenough and his team M. Stanley Whittingham, and Akira Yoshino around 1980. They received the Nobel Prize for their work in developing this battery. Since then much work has been completed in developing other ion batteries. The diagrams below are of a typical ion battery. They show the ion flow during discharge i.e.: when the battery is supplying power so the ions are moving from the anode (negative) to the cathode (positive) and the ion flow when the battery is being charged where they move back from the cathode (positive) to the anode (negative).

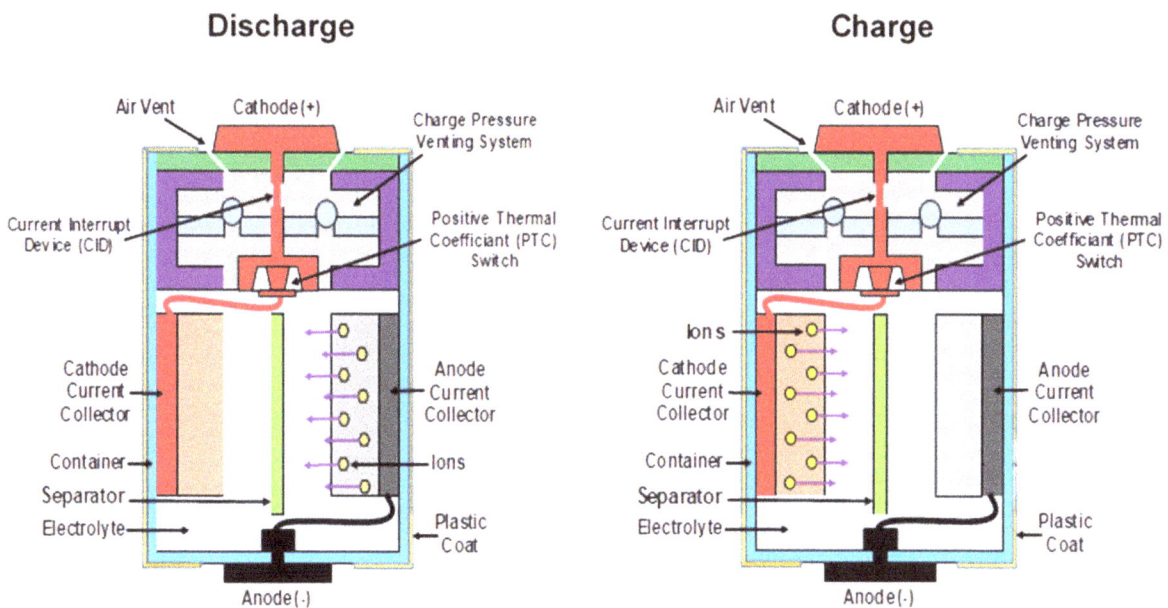

Discharge

Air Vent   Cathode(+)
Charge Pressure Venting System
Current Interrupt Device (CID)
Positive Thermal Coefficient (PTC) Switch
Cathode Current Collector
Anode Current Collector
Container
Separator
Electrolyte
Ions
Plastic Coat
Anode(-)

Charge

Air Vent   Cathode(+)
Charge Pressure Venting System
Current Interrupt Device (CID)
Positive Thermal Coefficient (PTC) Switch
Ions
Cathode Current Collector
Anode Current Collector
Container
Separator
Electrolyte
Plastic Coat
Anode(-)

Below I have shown several ion batteries that are not so common and are very unlikely to be in our Hazardous Area, but some may be batteries of the future.

1) Aluminium Ion Battery
2) Calcium Ion Battery
3) Carbon Ion battery (New Technology!)
4) Magnesium Ion Battery
5) Potassium Ion Battery
6) Sodium Ion Battery
7) Zinc Ion Battery
8) Hydrogen Ion Fuel Cell
9) Iron Ion Battery
10) Copper Ion Battery
11) Gold infused Lithium Ion Battery

Primary and Secondary Metal Air Batteries:

Air batteries are a type of fuel cell as they require a feed of external fuel to work. Usually the cathode is covered by a sticky cover and once removed the battery begins to work and they are self-discharging as they do not stop.

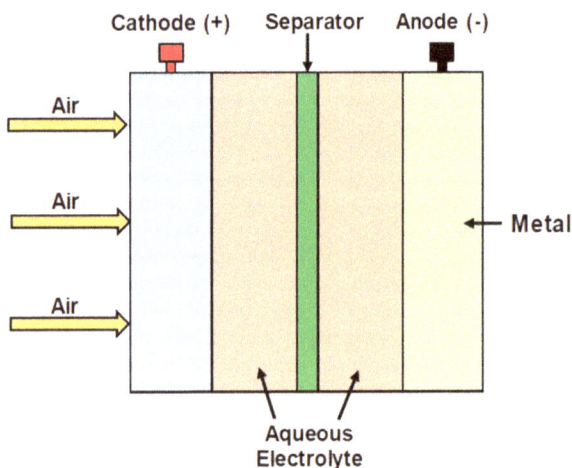

Left is a diagram of a metal air cell. These work by the cathode (+) being made up of carbon base, but mainly indrawn air, and the anode of some pure metal of which there are several different types listed below. The electrolyte is usually aqueous with a porous separator to stop any internal faults like the anode touching the cathode.

Making the cathode (+) of mainly air makes the battery much lighter. One popular air battery is the hearing aid battery (mentioned earlier). The moment you remove the plastic sticky cover you actually expose three very small holes which starts off the reaction. Voltages may be higher than 1.5 volts.

Remember not all air batteries are secondary. For instance in the Aluminium air battery the Aluminium anode degrades during the reaction and cannot be restored with a charge. The Zinc Air battery can be of primary or secondary design.

Some batteries are actually primary, but classed as secondary because you can actually replace the anode. Sodium air batteries are rechargeable but only for around 8 times. It is unlikely that you will find these in your Hazardous Area. Let us have a look at a list of several different types.

(**Note:** Voltages are approximate – several may vary!)

Primary Air Batteries:

1) Zinc Air Battery: (**Primary Type**) (1.4 Volts)
2) Aluminium Air Battery: (2.75 Volts)
3) Magnesium Air Fuel Cell: (3.1 Volts)

Secondary Air Batteries:

1) Zinc Air Battery: (Secondary Type) (1.4 Volts)
2) Germanium Air Battery: (1 Volt)
3) Calcium Air Battery: (3.1 Volts)
4) Iron Air Battery: (1.3 Volts)
5) Potassium-ion Air Battery: (2.5 Volts)
6) Silicon Air Battery: (1.2 Volts) Experimental!
7) Tin Air Battery: (1 Volt)
8) Sodium Air Battery: (2.3 Volts)
9) Beryllium Air Battery: (1.3 Volts)
10) Lithium Air Battery: (2.9 Volts)
11) Sugar Air Battery
12) Vanadium Boride Air Battery

Primary and Secondary Silver Batteries:

Below I have listed batteries containing Silver. As you can imagine these batteries may be expensive because of the cathode electrode metal.

1) Silver Zinc Battery (Primary)

2) Silver Zinc Battery (Secondary)

3) Silver Copper Battery

4) Silver Cadmium Battery

5) Silver Calcium Battery

Primary Zinc Silver Oxide Battery:

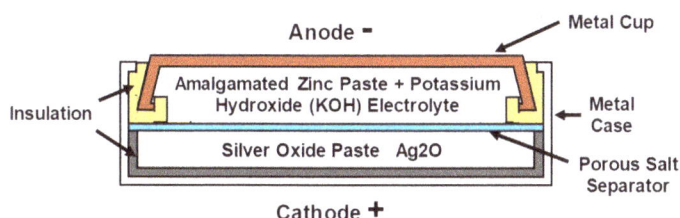

We have already discussed these batteries. They are **PRIMARY BATTERIES** which means that they **CANNOT BE RECHARGED.** Its flat shape sometimes earns it the name of a 'Button' Battery. Used as a power source for equipment such as watches, hearing aids etc.

Although expensive these small batteries tend to give a long life. They have Silver as part of the chemistry which makes them expensive. Usually they have a finite life, but it is possible to obtain a rechargeable version of this battery. These may be confined to instances where they are charged by the device and have to hold information even if the device is switched off or the main device battery has gone flat.

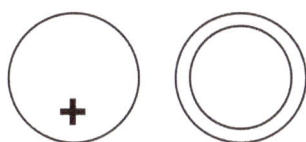

Some of these Batteries tend to be made using Silver Oxide (Ag_2O) + Zinc (Zn). Silver Oxide batteries may be expensive because of what it consists of in its chemistry. Other materials can be used in these 'Button' batteries as below.

At the top of the page is a diagram of a Silver Oxide 'Button' battery just to give you an idea of the construction, but there are several other similar batteries with different anode materials such as Lithium (Li) instead of Zinc (Zn). Lithium batteries are discussed under their own section in this book. Some different cathode materials can be: Copper (Cupric) Oxide (CuO), Manganese Dioxide (MnO_2) or Carbon Monofluoride (CS) which of course is an alkali. Something else that unusually can be used as a cathode material is Oxygen (O_2). Certain hearing aid batteries in the past used to use one material called Mercuric Oxide (HgO) but as you can imagine in an environmental friendly world anything containing Mercury (Hg) would not be used in modern times. Instead of Mercury (Hg) the manufacturers may use Lead (Pb) which in itself is not the most environmental friendly element.

The M and N sizes were 'Mercuric Oxide' and are not used anymore. Again the L, S, P, G and Z batteries are the standard **1.5 volts.** The C and B sizes are **3 volts** so it is very important that the correct size and voltage is chosen. A typical calculator battery may have the number: **CR2025.** The **first letter** denotes the type i.e.: **C** = Lithium, **S** = Silver Oxide, **P** = Zinc/Oxygen, **L** = Manganese Dioxide, **B** = Lithium Carbon Monofluoride, **G** = Lithium Copper Oxide and **Z** = Nickel Oxyhydroxide. The **second letter** denotes the **shape**: **R** = Round, **F** = Flat, **S** = Square and **P** = the rest of the shapes not mentioned! The numbers denote the primary element/chemical (i.e.: Lithium), thickness of the battery and the diameter. **So our Battery CR2025 = Lithium Round 20mm diameter and 2.5mm high.**

One of the main safety problems with this type of battery is that it is so small that it can be easily swallowed by children. IF A BATTERY IS SWALLOWED BY A CHILD IT CAN BE FATAL. YOU MUST TAKE TO ACCIDENT AND EMERGENCY IMMEDIATELY! IT MAY LODGE IN THE STOMACH OR INTESTINE AND SOME ARE VERY, VERY DANGEROUSLY CORROSIVE AND/OR TOXIC.

Secondary Silver Zinc Battery:

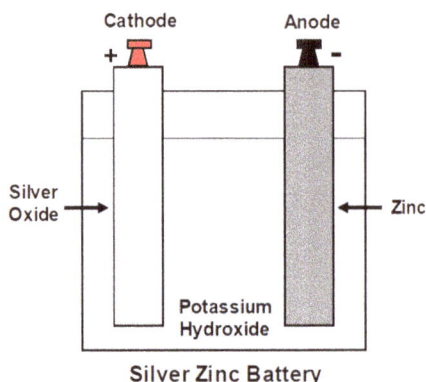

Cathode   Anode

Silver Oxide →   ← Zinc

Potassium Hydroxide

Silver Zinc Battery

So with a cathode of Silver Oxide (AG_2O) and an anode of Zinc Oxide (ZnO) in an alkaline electrolyte of Potassium or Sodium Hydroxide, we have our battery. Enjoys a high energy density. The voltage of these batteries is around 1.5 volts. The cost as you might expect is high, but the battery life is around 3–5 years.

On discharge Zinc in the anode becomes Zinc Oxide and the Silver Oxide cathode becomes pure Silver so recycling is vital at the end of battery life. When stored, self-discharge is around 4%/year. Uses are mainly military and space.

Secondary Silver Copper Battery:

This will be called a **Redox** reaction battery. I have drawn the diagram as two half Galvanic Cells. The half-cell on the right contains an electrode of Copper (CU) in a Copper Nitrate Solution ($CU(NO_3)_2$). The Copper anode is oxidising. The half-cell on the left contains an electrode of Silver (Ag) in a Silver Nitrate Solution ($AgNO_2$). The silver cathode is undergoing Reduction. The Salt bridge of Potassium Nitrate (KNO_3) is to allow the anions and cations (cathode) to pass onto the electrodes. (Ion Flow).

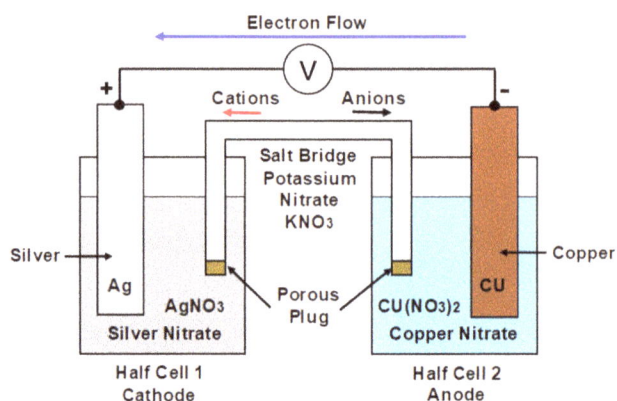

Electron Flow

Cations   Anions

Salt Bridge Potassium Nitrate KNO_3

Silver — Ag $AgNO_3$ Silver Nitrate   Porous Plug $CU(NO_3)_2$ Copper Nitrate   CU — Copper

Half Cell 1 Cathode   Half Cell 2 Anode

Secondary Silver Cadmium Battery:

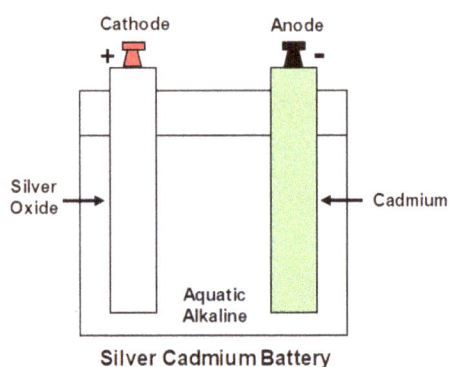

Cathode   Anode

Silver Oxide →   ← Cadmium

Aquatic Alkaline

Silver Cadmium Battery

As per the diagram on the left these batteries have a Cadmium (Cd) anode (negative) electrode and a Silver Oxide ($AgNO_3$) cathode (positive) electrode with an Aqueous Alkaline electrolyte. Around 1.1 volts per cell which is lower than other batteries. More expensive against, say, the NiCd Battery because of the Silver but more energy productive. Cadmium is toxic so its environmental status would not be as good as other batteries. Separator in the electrolyte to stop the anode touching the cathode and also to prevent migration of the Silver Oxide. Transported dry and electrolyte added by the user.

Secondary Silver Calcium Battery:

These are very similar to the **Lead Acid** battery mentioned earlier. **Battery Grid Make up:** Lead, Calcium and Silver Alloy (instead of Lead Antimony Alloy). **Charging: Voltage:** is slightly higher than lead acid around 14.8 volts. **Battery Sulphation:** is a problem here where crystals coat the plates and the battery becomes inefficient. **Electrolyte Stratification:** again is a problem here where water in the electrolyte separates from the acid and rises to the top of the battery with the acid at the bottom. This of course will lead to the battery efficiency dropping.

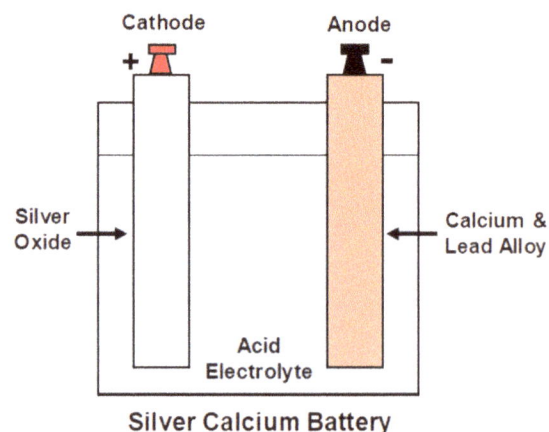

Cathode   Anode

Silver Oxide →   ← Calcium & Lead Alloy

Acid Electrolyte

Silver Calcium Battery

Other less Common Batteries:

Some of the following batteries I am sure that you will have never heard of and after reading this book will never hear of again! Although you may not find them in your Hazardous Area they do exist and sometimes you get the idea that some of these batteries may be the ones of the future with a bit more research. Some of them are extinct and some will be in the near future.

Secondary Carnot Battery:

These batteries are very unusual as they do not use the chemical reaction to produce electricity, they use heat reaction. When the battery is being charged it used the charger electricity input to produce heat in the first 'exchanger' which is stored in the 'heat store' as heat energy in the middle unit and when the battery is being used it releases that heat energy through the second exchanger to produce electricity. Could this be a battery of the future? It sounds very environmental friendly!

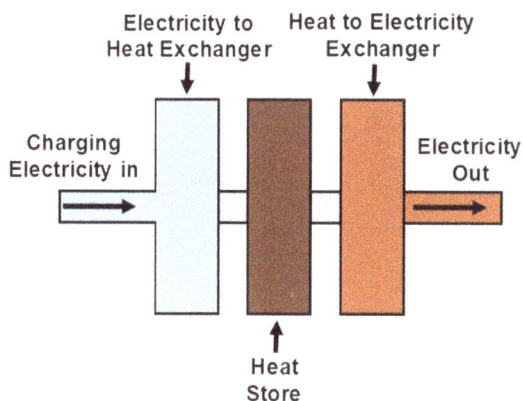

Primary CMOS Battery:

CMOS stands for **C**omplementary **M**etal-**O**xide **S**emiconductor. These are used on PC equipment motherboards to maintain important information even if the device is switched off or its power battery flattens. Very long life! CMOS should not be handled as even the static charge in your body no matter how small can damage.

Glass Battery:

A glass battery is not what it appears to be by the name. Glass is actually the electrolyte, usually with Lithium Metal electrodes. Again this battery was invented by John Goodenough and Maria H. Braga.

Molten Salt Battery:

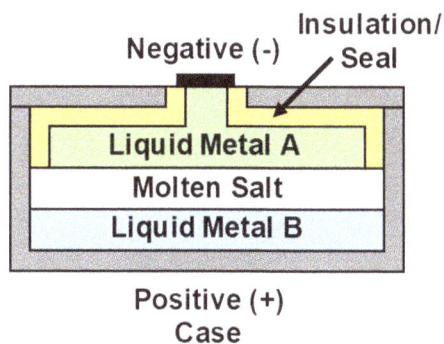

These batteries, as per the sketch to the left, use two molten metals as the anode (negative) and cathode (positive) separated by a molten salt electrolyte.

Nano Battery:

This is a battery using **Nano Technology.** If we put a coating of Nano Particles on the surface of an electrode of this battery we actually increase the surface area of that electrode making it more efficient. Nano technology is the ability to directly control each atom in a material. This is very sophisticated and future technology.

Secondary Organic Radical Battery:

Sometimes called a **Polymer Based** battery. This battery design is not like any conventional batteries that we have come across. Many batteries use dissimilar metals to create a voltage, but this battery uses polymers as in the sketch to the right. The cathode (positive) can be Nitroxide Oxoammonium and the anode (negative) Nitroxide Hydroxylamine polymers. This is a secondary battery so can be charged.

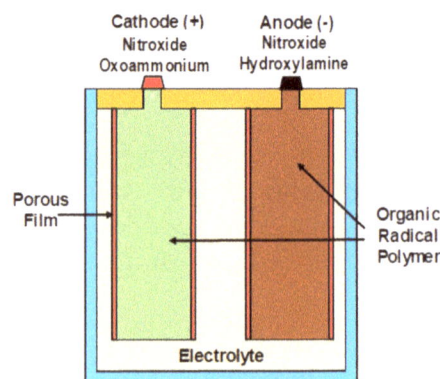

Primary Photoflash Battery:

These are a battery of the past as separately powered flashes are no longer required. Basically these are a Zinc-Carbon battery slightly modified to give one burst of high voltage and current power to the flash unit when required.

Primary Pola-pulse Battery:

These Zinc Chloride batteries as above were batteries of the past and anyone my age will remember a Polaroid Camera which took pictures and developed them inside of the camera and out they came. This battery was specially designed for these cameras.

Primary Microbial Fuel Cells:

These are called **'Fuel Cells'** rather than batteries and produce electricity by using bacteria and Oxygen (O_2). The very first prototypes caused electrons to be transferred from the bacteria to the anode using a chemical. (This was called a **'Mediated'** process.) The other process is what is called an **'Unmediated'** which came out later and electrons were transferred direct to the anode. These are energy producers of the future, but at the moment are very rare with limited uses.

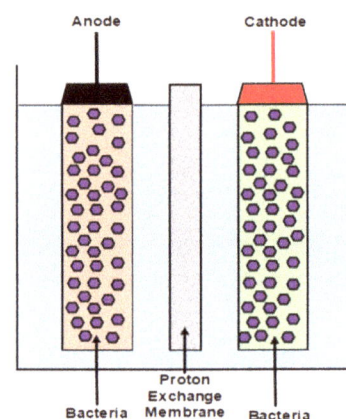

Sodium Sulphur Battery:

This battery has an operating temperature of around 340C so it is very hot! Again there are similarities to the Molten Salt battery as it uses two liquid salts i.e.: a Sodium and Sulphur type. One of the problems they had to overcome was that the actual contents of the battery were very highly corrosive, as we have found out with Sodium lighting. Sodium and damp air are explosive so the contents must never be allowed out of the battery.

Secondary Solid State Battery:

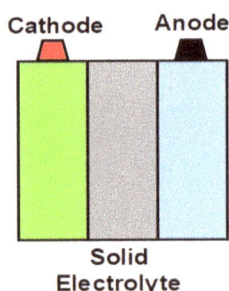

These batteries are called a **'Solid State Battery'** because of the solid electrolyte whereas in most batteries this would be a gel, paste or liquid. Nothing to do with electronics. The main problem with these batteries seems to be cost, they can be very expensive. The solid electrolyte seems to be the stumbling block. Several materials used are glass, Sulphides & polymers. Lithium may be an important part of these batteries in the future and much development is being completed. They are environmentally friendly.

Sugar Battery Experimental:

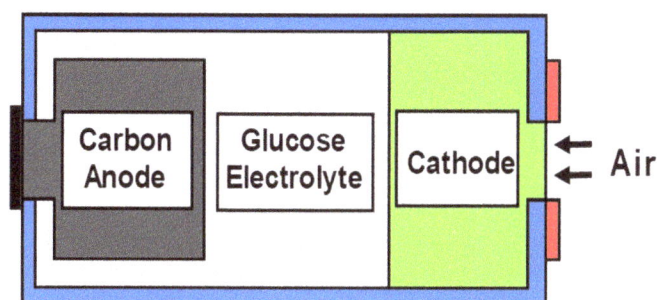

Humans discovered that glucose gives you energy long ago. Can we tap this energy source into a battery? Well I have shown this as almost a home-made battery but serious research/manufacture is close. This new **Sugar Battery** is in its embryo stage, but by using a Carbon based anode (-), an **Air Breathing** cathode (+) and a Glucose electrolyte, hopefully they will end up with a very high powered environmentally friendly battery.

Super Iron Battery:

Cathode Potassium Ferrate (K_2FeO_4) or Barium Ferrate ($BaFeO_4$) instead of Manganese Dioxide (MnO_2). Because the chemicals have an Iron (Fe) content the depleted cathode is mainly rust (Iron Oxide (Fe_2O_3)). Electrolyte Potassium Hydroxide (KOH). The anode is a Zinc metal alloy. Large increase of around 50% capacity on rival batteries. Battery under massive development. At present looking at AA and AAA Sizes.

Ultra Battery:

This battery to me is a super battery of the future. It is an offshoot of the Lead (Pb) Acid (H_2SO_4) battery. Built into the battery is an 'Ultra Capacitor' (Supercapacitor). We end up here with a battery that can be charged very fast. Its resistance to sulphation and above fast charge make it ideal for hybrid electric vehicles. A bit doubtful for pure electric vehicles though. Does Lead let it down as far as being environmentally friendly goes?

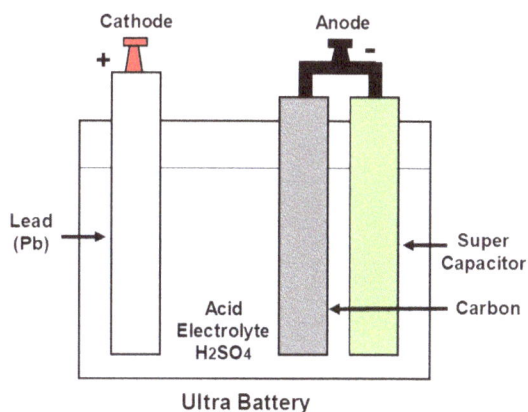

Ultra Battery

Wiltshire Battery Power:

Projects like the one in Minety, Wiltshire may be our energy supplies of the future. Pensa Power have built a battery as large as a football field, the largest in Europe, which is charged by environmentally friendly renewable energy such as wind farms and solar energy etc. and feeds onto the National Grid. The system is 100 megawatt.

The site is unobtrusive and with no adverse effect on the countryside. It is hoped that many of these facilities will be built all over the country giving us future reliable power.

Future Batteries:

They are on the verge of bringing out a battery that will power a device such as a smart phone that will charge in seconds and last a considerable amount of time. The battery is likely to be Lithium but a lot of research is being done on other types of battery such as Sulphur, sugar and many more. They are very environmentally friendly so watch this space!

I am sure that they are also on the verge of bringing out a car battery that again will have a very fast charge, possibly minutes, and will again last much longer than the ones at present so again watch this space.

Several Home-made Primary Batteries:

Primary Earth Battery:

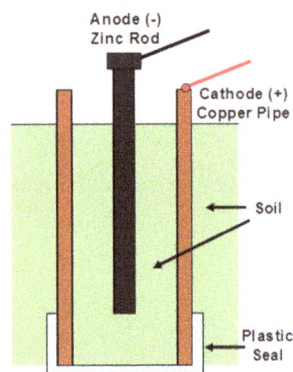

Earth Batteries can be made very easily with a length of Copper pipe filled with soil and sealed at the bottom with, say, a plastic material over the end before inserting it into the ground. A Zinc rod is then inserted into the soil inside of the copper pipe down into the soil, as in the diagram on the left. The Zinc rod then becomes the anode and the Copper pipe in the ground in fact ends up as the cathode. As you may have gathered the damp soil inside of the Copper pipe acts as the electrolyte between the two dissimilar metals. The type of metals used will determine what voltage is produced at the anode and cathode. Zinc, as in the diagram, may produce around 1 volt. The metals can be put into the ground separately without using a pipe.

Lemon Battery:

The **Lemon Battery** is simple to make. Half a lemon with a Zinc plate and a Copper plate connected to a LED. The EMF is extremely small and it may require several lemon batteries in series for the experiment. Again here we have two dissimilar metals in an acid type electrolyte. I have chosen a lemon but this experiment should work with any acidic fruit and different metal spikes or nails hammered into a lemon. The same experiment can be completed using a potato. I have done the lemon experiment and it does work but very low EMF.

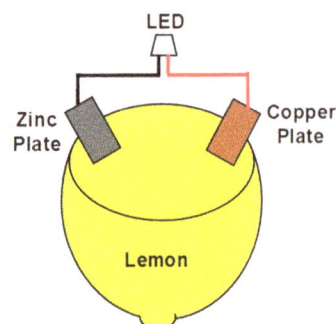

Penny Battery:

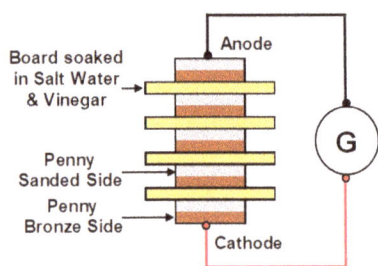

For the **Penny Battery** take 5 or 6 pennies and gently file one side until you see silver Zinc as shown in the diagram left. Place a piece of thin card between them damp with salt water & vinegar. This is the electrolyte. Stack one on top of the other remembering to keep the coins the same way up with the damp card separating them. Connect to a **Galvanometer** or very low current device. An EMF should flow from the two dissimilar metals. In an analogy this would be similar to a thermocouple.

Paper Battery:

The majority of batteries are made up of two dissimilar elements, sometimes metals, in an electrolyte and the result is a potential EMF. This battery is no different, there is a vinegar soaked paper electrolyte separating two electrodes. One electrode, the anode (-) is Aluminium and the other electrode, the cathode (+) is Copper. These two elements are commonly used in many batteries.

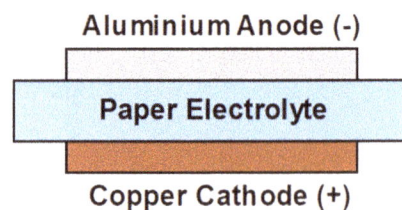

Salt Water Battery:

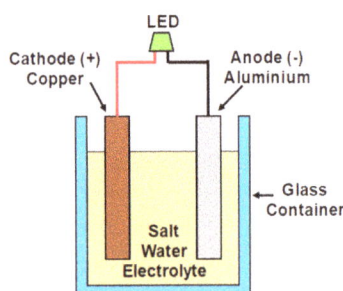

As in manufacture the majority of batteries are made up of two dissimilar elements, sometimes metals, in an electrolyte and the result is a potential EMF. This battery is no different, the **Salt Water Battery** is another simple environmentally friendly battery. The idea is that a container is filled with salt water. This is the electrolyte. Placed in the container is a Copper electrode which we will call the cathode (+) and an Aluminium electrode which we will call the anode (-) The set up in the diagram on the left will just about light an LED, but larger batteries can be constructed.

Battery Banks:

When designing or thinking of acquiring a battery bank there are several factors that we must look into as regards the batteries themselves, namely **Voltage (V), Amp hours (Ah), Series** and **Parallel.** It might be the Amp hours that may give us problems. Batteries in series as below may be called a **'String Voltage'**. Let us have a look at some basics:

1 – Three batteries of the same voltage in series:

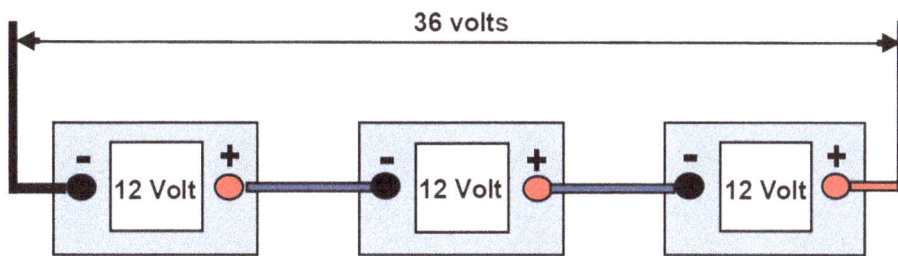

2 – Three batteries of different voltages in series:

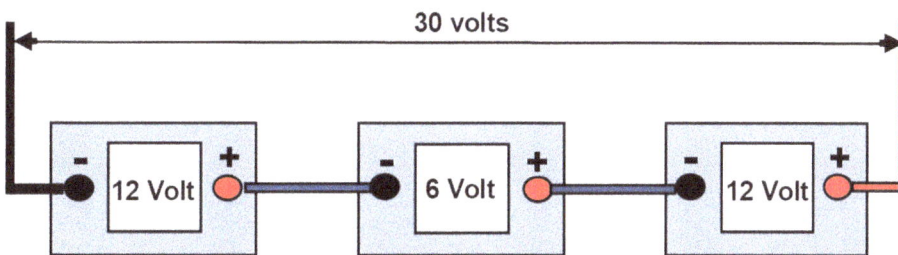

Not a great deal of a problem here when we are just looking at the discharge voltage. So, in theory, we add all of the voltages together which equals **30 volts** output. All you have to watch here is that some **12 volt/6 volt** batteries actually vary slightly in voltage and may not be exactly **12 volts/6 volts** but this can only be looked into by measuring the voltage of each battery with a voltmeter.

Charging may be the biggest problem. You would have to obtain data from manufacturers etc. about the action of connecting a 30+ volt charger to the output and the effect it would that have on the different voltage batteries.

3 – Three batteries of the same voltage and Amp hours in series:

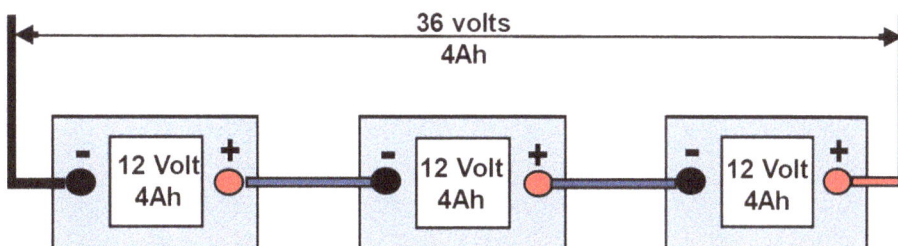

Not a great deal wrong here as the voltages are the same and the Amp hours (Ah) are the same so we end up with a small bank connected in series so adding the voltages up comes to **36 volts.** The Amp hours will remain at **4Ah**. So you can supply **4 amps** for **1 hour** at **36 volts. Again charging would have to be looked into.**

Note:

As we go through this section we will look at batteries in series and parallel with all of the dos and don'ts which might innocently be carried out on a battery bank individually designed without the manufacturers being involved.

We will be looking in detail how we can easily fall fowl of voltage and Amp hours and see what happens if we get it wrong even with primary batteries. Lastly we will discuss why we would connect a battery bank in series **AND** parallel to obtain the right voltage and Amp hours for our UPS system.

4 – Three batteries of the same voltage with different Amp hours in series:

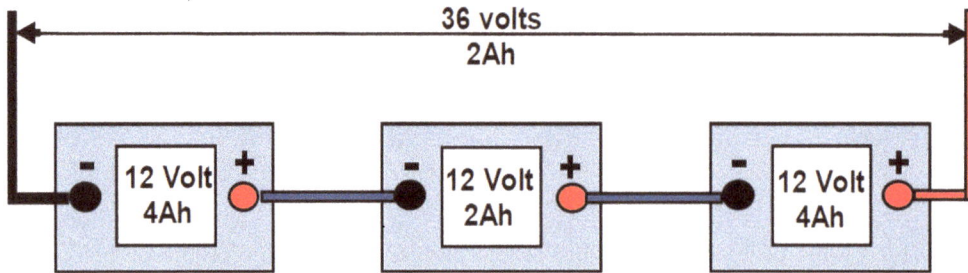

Here is where you have to be careful. This time there are 3 batteries with the same voltage which is no problem at all. In series the above set up will give you **36 volts**. The main problem is that the middle battery is only **2 Amp hours** whilst the other two are **4 Amp hours**.

In reality the middle battery at **2 Amp hours** will run flat sooner than the other two which are **4 Amp hours** so the whole bank must be gauged on the weakest battery. So, going by the latter, you would end up here at **36 volts at 2 Amp hours**.

What you also have to remember here is that the lower Amp hour battery could go completely flat well before the others, which might not do the battery any good. As the other two will still be trying to supply the load, will their current rise beyond tolerance? If so the extra current could cause severe damage inside of the batteries. Manufacturer's design engineers should be consulted in this instance.

Also another problem here is charging the batteries. What will we actually achieve by connecting the charger to the output? If we charge up the **2 Amp** hour battery very quickly, will the other **4 Amp** hour batteries take much longer? If so the result would be overcharging the one 2 Amp hour battery whilst the others become fully charged which, again, could cause damage.

5 – Three batteries of the same voltage in parallel:

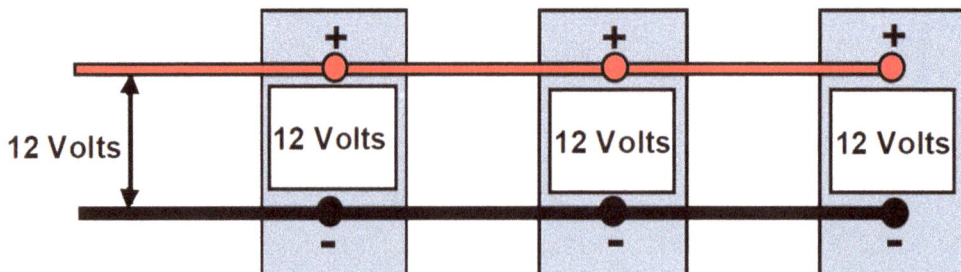

In the diagram above we have 3 batteries connected in parallel. No problems here with the voltage, it will remain the same as the battery voltage of **12 volts** which is too low to do much good for a UPS.

Of course when you use jump leads on your car this is exactly what you are doing with just 2 batteries. Batteries are put into parallel like the 3 batteries above so that their **'battery run time' (Ah)** will last much longer as we will see later in this section.

Note:

As we go through this section we will look at batteries in series and parallel with all of the dos and don'ts which might be innocently be carried out on a battery bank individually designed without the manufacturers being involved.

We will be looking in detail how we can easily fall foul of voltage and Amp hours and see what happens if we get it wrong even with primary batteries. Lastly we will discuss why we would connect a battery bank in series **AND** parallel to obtain the right voltage and Amp hours for our UPS system.

6 – Three batteries, one of lower voltage in parallel:

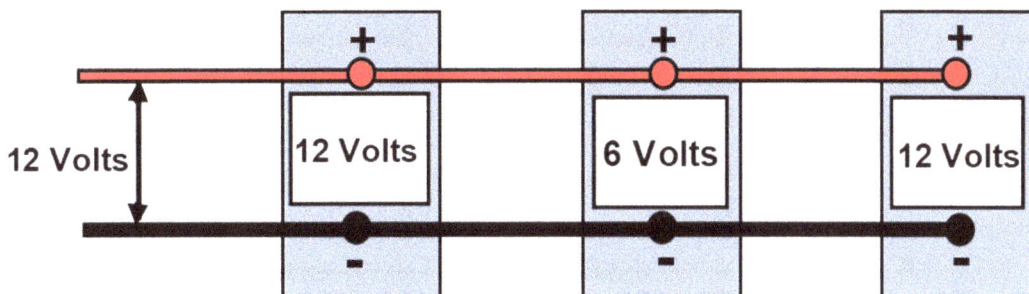

This is one big don't do! If you connect batteries of different voltages in parallel, what will happen is that the batteries with the higher voltages i.e.: **12 volts,** will be constantly trying to charge the battery with the lower voltage i.e.: **6 volts** past its charge absorption point to try and bring about equilibrium. The **6 volt** battery could overheat and damage occur inside of the battery or worse.

I know we are talking about secondary batteries, but mixing primary batteries of different makes and/or types can bring about the same thing. Some primary batteries are 1.5 volts, some are 1.7 volts and mixing them together in equipment will bring about what we have been describing. Batteries can explode!

7 – Three batteries, all of different Amp hour in parallel:

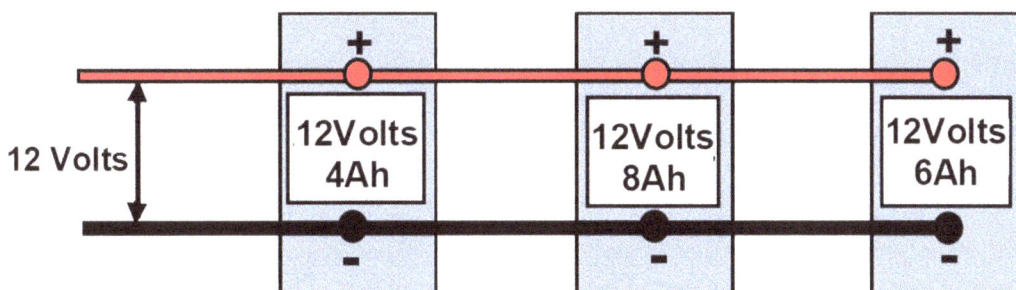

Just got to be careful doing this because of charging differences. The voltage will remain at **12 volts**. The Amp hours however are added together so we have **4 Amp hours + 8 Amp hours + 6 Amp hours** which total **18 Amp hours.** Therefore we have an output of **12 volt 18 Amp hours.**

Try to ensure that the batteries are of the same make & type and try **NOT** to have the Amp hours (Ah) different otherwise other issues come into play such as charging voltages and times. For instance will the **4 Amp hour** battery charge much quicker than the other two so by the time that **8 Amp hour** battery is charged, the others be overcharged?

It is always wise in any situation to ensure that all of the batteries in the bank are identical in make, type, voltage and Amp hour.

We have now got to consider how we get enough voltage to feed our UPS with as many Amp hours as possible. We have looked at a series connected set of batteries where we add all of the voltages up from each battery and now we have looked at a parallel connected set of batteries where we add up the Amp hours. Is there a way of connecting a battery bank so that we can get the benefit of both series and parallel connections?

71

8 – Nine batteries in series AND parallel:

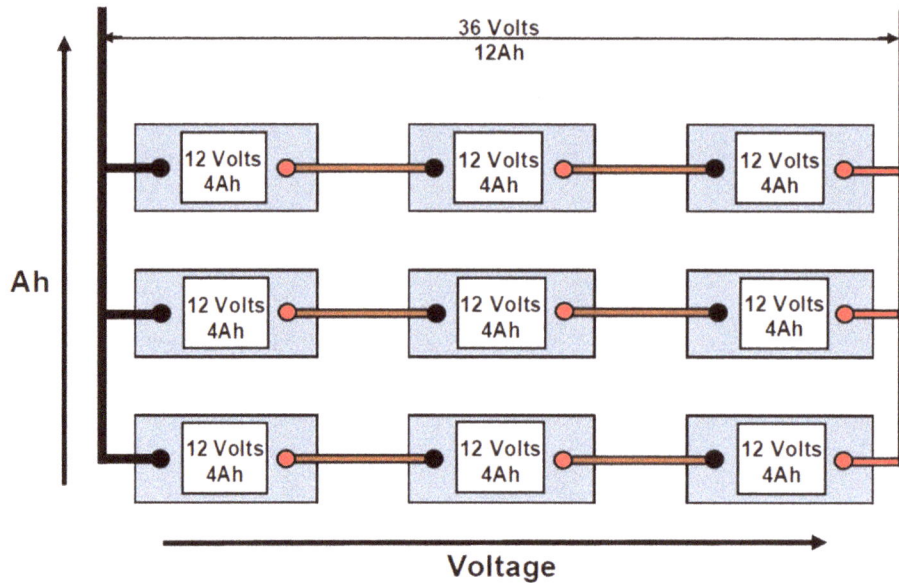

We talked earlier about getting the better of both worlds i.e.: series and parallel. We said that batteries in series gave us higher voltage and batteries in parallel gave us higher Amp hours, so what if we combine the two into our battery bank?

Firstly let us look at the **voltage.** In the diagram above the batteries are connected in series **AND** parallel at the same time. In series (cross ways) we add the voltages together so in the above diagram we have three sets of batteries each of them **12 volts** in series which comes to **36 volts** per set. Parallel the 3 sets up and we still have **36 volts**. Now we come to the Amp hours which if you remember we add up so looking at the diagram again we have 3 sets of batteries in parallel and the batteries in each set (vertical) are **4 Amp** hours. We now add the Amp hours in the line which comes to **3 x 4Ah = 12 Amp hours.**

So to sum up we have **3 sets** of **3 batteries**, each of which is **12 volts**, in series which comes in our case to **3 x 12 = 36 volts**. We also have **3 sets** of batteries in parallel **3 x 4 = 12 Amp hours**. So to build up our bank we put as many batteries in the line as it takes to make our output in volts so if we required **240 volts DC** that would be **20 batteries** in our series line **20 x 12 = 240 volts DC** and how many Amp hours we required would depend upon how many series lines or what size of each battery, in Amp hours, of **20 batteries in parallel**. The larger the Amp hours the fewer battery series lines.

Let me just say here that this section is just a tutorial of how battery banks may be connected and expert advice should be sought in the design of battery banks.

Several factors must be met. Each battery must be:

1) The same **voltage**. If not, the other batteries will be constantly trying to charge those of lower voltage to balance the system.

2) The same **voltage**, otherwise the charger voltage will overcharge any lesser voltage batteries and heat and damage will be produced.

3) The same **Amp hour**. Again the charger will charge any batteries with lower Amp hours very quickly and heat and damage will be produced.

4) The same **Amp hours** or any batteries that are fewer Amp hours may go completely flat which could affect the overall battery voltage in the bank.

Sometimes designing the battery supply for, say, a UPS emergency lighting system is not as easy as it may seem. The battery voltage is DC (Direct Current) for a start, but our emergency lighting is AC (Alternating Current).

It may not be as easy as just putting our 240 volt DC supply into an inverter and getting 240 volt AC out. The inverter circuit itself will require a certain amount of power, the batteries may have a slight tolerance in their voltage etc.

Remember for best efficiency and battery life, every battery in the system must be absolutely identical and manufacturer design and advice followed.

9 – UPS requiring 240Volts at 100Amp hours:

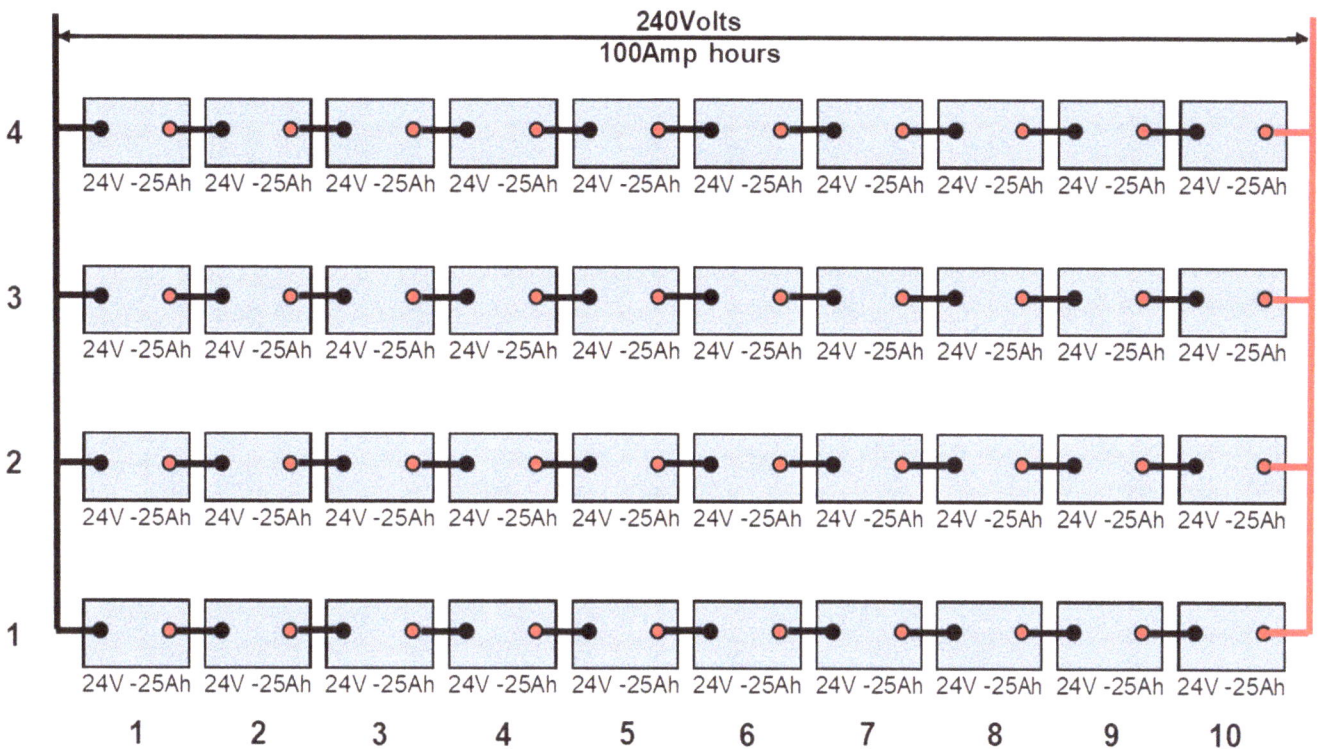

Above is a simplified diagram of the configuration of a battery bank to give **240 volts (1–10 horizontal)** at **100 Amp hours (1–4 vertical)**. Each of the 40 batteries in the bank is **24 volt 25 Amp** hours. Let us use the emergency lighting UPS as an example. Output required depends largely upon:

1) How many lights are on our emergency system (Amperage)? To obtain estimated current.

2) What type of lights are they? (Fluorescent, LED etc.)

3) How long, say in hours, would you want them to stay on for in a power outage?

4) What are the expected manufacturer voltage tolerances per battery? (Manufacturer's Spec.)

5) Is there any voltage drop involved caused by resistance in long cables?

6) How much power does the inverter circuit itself require when running? (Manufacturer's Spec.)

7) How often are the batteries maintained? (Company Maintenance System)

8) The batteries must not go into a stage where they are completely flat otherwise this would be detrimental to their life. Manufacturers will have a certain minimum voltage/battery which the UPS Inverter cut out should not drop below.

9) Not exactly the output, but what is the charging system that is used?

Battery Banks that feed UPS Systems can be known as 'BESS' (Battery Energy Storage System).

UPS System:

UPS Systems are used on Chemical Factories for a number of different things. On our plant we had two different inverters and a 24V DC Supply:

1) **The Emergency Lighting Inverter:** This fed normal **240 volt AC** fluorescent lighting on the plant. So in a power outage the inverter would kick in and supply power to the emergency lights for some considerable time. Switching can be achieved by an electronic static switch which provides the emergency power when required.

   The lighting inverter can be tested once a month for, say, 2 hours by manually operating the static switch. Operators would check the lighting for faulty fluorescent tubes/fittings at the same time. System cut off would ensure that if the lighting was left on to the battery maximum time, then the batteries did not drop below the minimum volts per cell.

2) **The Instrument Inverter:** This would supply vital instruments on plant with **110 volt AC** on the plant. Again in a power outage the inverter would kick via a static switch and feed vital instruments which would be required to inform process of the plant condition even in the event of a main power cut. A bit more care to be taken testing this inverter as it could send the plant into a crash shutdown.

3) **24 volt DC Supply:** Again vital. **24 volt DC** instruments would have to be fed from this supply. No inverter involved here just a battery bank, static switch and charger.

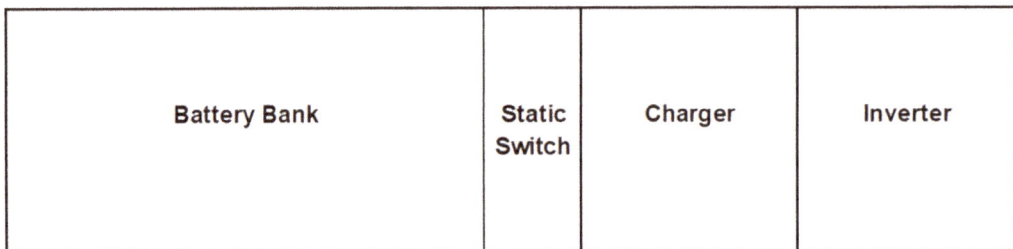

A small UPS may look similar to the diagram above where the batteries are integral to the Inverter and charger. In other cases the plant would have a 'battery room' which was full of batteries which may be supplying the emergency lighting system.

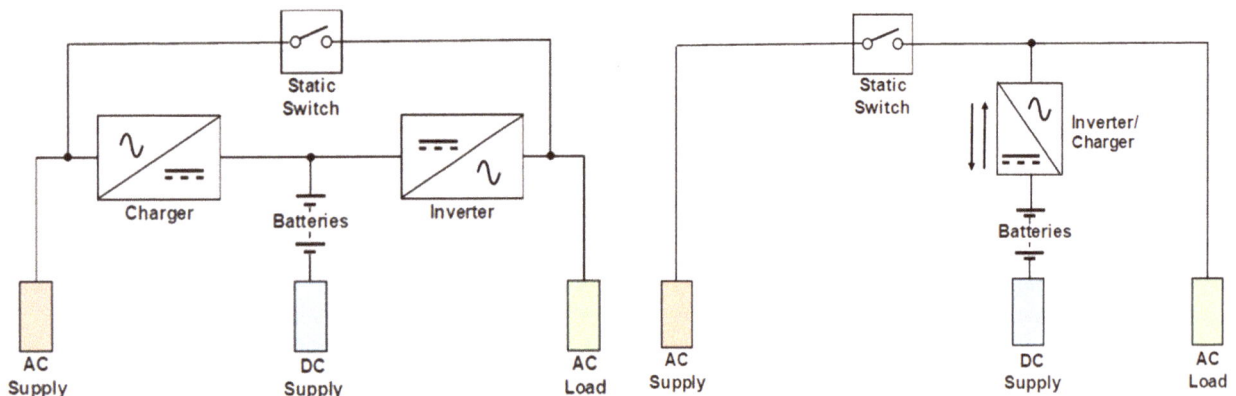

The UPS systems have many different configurations. I have shown two different systems that we had. Above left has a **separate charger** and **overall static switch.** So the charger charges the batteries whilst the inverter is not required and when the static switch operates under supply failure, the inverter takes over.

Above right shows a **bi-directional** inverter which has an integral battery charger which charges the batteries whilst everything is normal and the inverter will take over in the result of a power loss. There are advantages and disadvantages in both systems and it is always advisory to get manufacturer's guidance when selecting a system to suit your needs.

Solar Cells:

Let us have a look at a very basic explanation of how solar cells work and how they are connected to form useful energy in the form of electricity from sunlight.

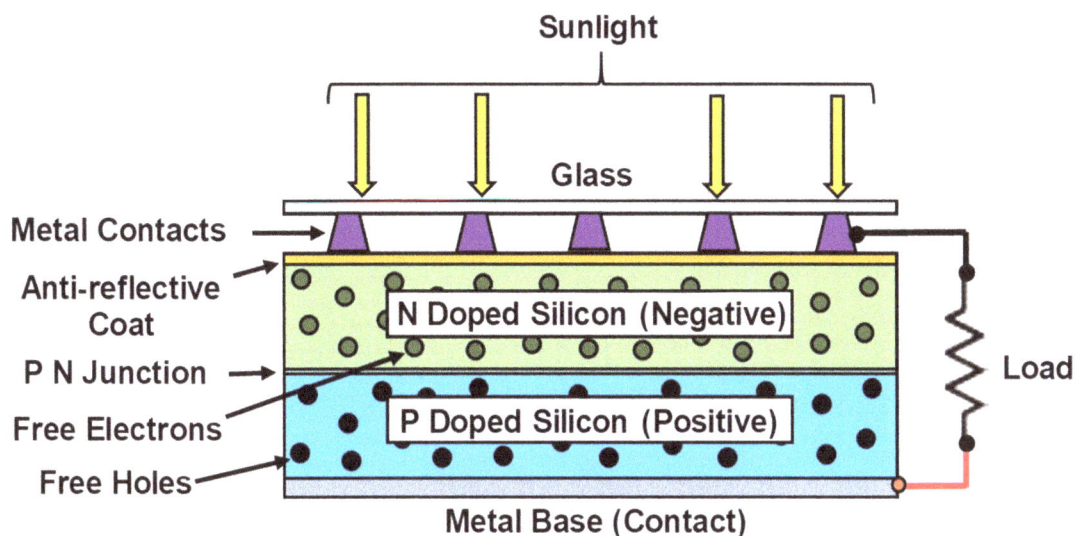

Silicon is one of the most abundant materials on earth. In terms of sand, we have deserts with billions of tons of it and with a little alteration, such as the addition of Carbon and Hydrogen, can go through a process to convert it into almost pure silicon. Sunlight is also available as free energy. Can we bring these two abundant commodities together to form another type of energy, namely electricity?

Above we have a diagram of a solar cell or **photovoltaic (PV) cell.** There are two semi-conductor compounds of Silicon, in a sandwich is the best way to describe it. Impurities are added to the silicon, called doping, to make it a better conductor. The top layer of Silicon (green) is doped with Phosphorous and in doing so makes it negative and the bottom layer of Silicon (blue) is doped with Boron making it positive.

So to sum up, when sunlight **'photons'** hit the top glass, the top layer of Silicon (green) via the metal conductors and anti-reflective coat into the N doped silicon layer, how does this cause electrons to flow? Well by doping the Silicon negative layer (green) with Phosphorous we can make the electrons in the Silicon much looser.

By doping the Silicon positive layer (blue) with Boron we can create 'holes' around the atoms. Put the two together with a PN junction and some electrons from the negative will fill the holes in the positive and some of the holes will migrate to the negative.

As photons penetrate the negative Silicon layer, an electric field will be formed in the PN region. Holes will flow from the positive Silicon (blue) to the negative Silicon (green) and electrons vice-versa so in actual fact a potential difference is formed.

Now if we connect a load as in the diagram above, electrons will flow through the load just the same as if we had connected it to a battery. Although in the diagram for ease of explanation I have made the two Silicon layers of equal thickness, in reality the negative (green) layer will be thinner than the positive (blue) layer.

So now we have our solar battery working with electron flow, but at the moment with only one cell the amount of energy is so tiny that it is of no use to us. So how are we going to magnify this tiny amount of energy into something we can use?

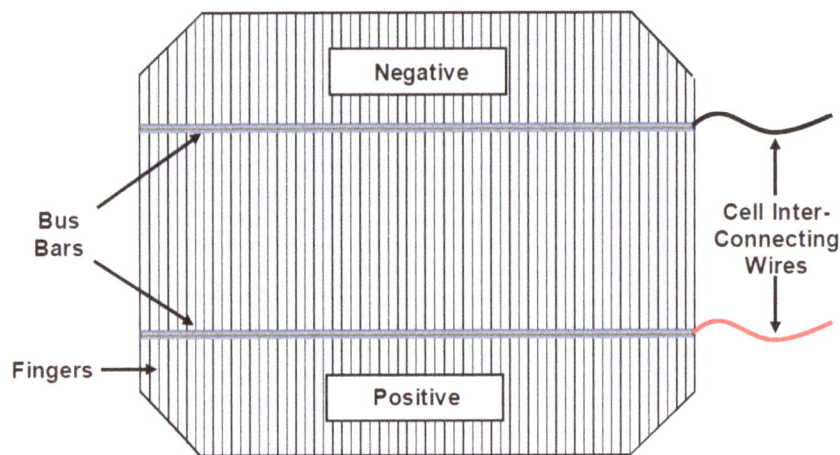

Above is the make-up of a single solar cell and electrons pass through the fingers and collect on the two bus-bars. As mentioned previously we have a very small photovoltaic cell, but the electrical energy is so minute (say half a volt) that alone it would be of no use to us. So what is our next step?

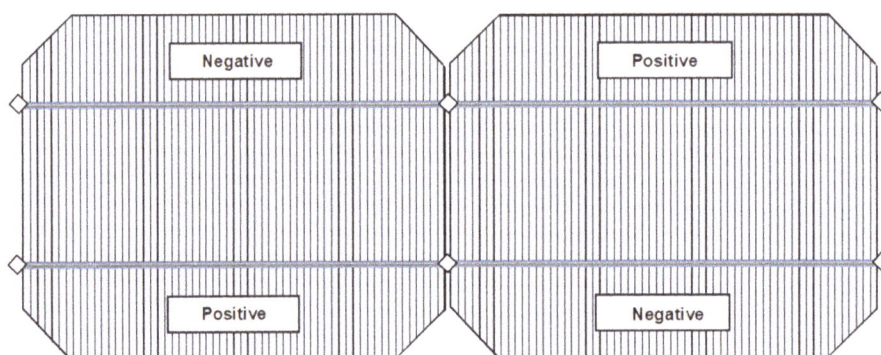

Well as you can see in the diagram above we have connected the cells in series so from what we have learned so far the series circuit causes a rise in voltage, but still no real usable power. We now connect the cells together to make solar modules, but we are not finished yet.

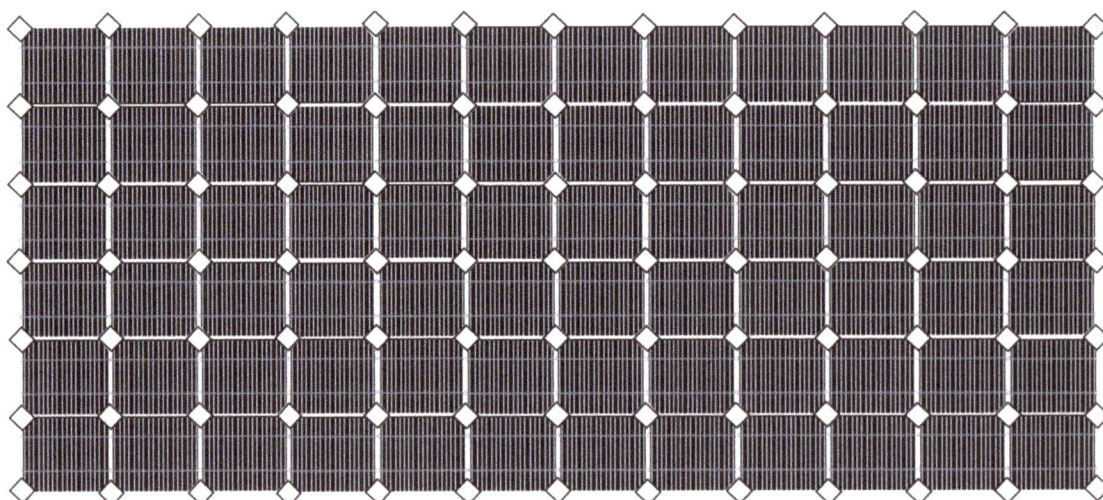

Above we have connected a number of cells (72) together in a series/parallel formation to make a solar panel. If you remember from our batteries, series lifts the voltage and parallel lifts the current (Ah). So these are what you see in numbers on house & factory roofs. We can also put many panels in a Solar Panel Farm in a field and make what is called a Solar Array.

Solar Panels:

Sometimes some solar panels look totally different from others. This is because of the way the crystal lattice is formed within the panel.

1) The purple coloured panels which probably are the most common are called **'Polycrystalline'** **Cells.**

2) The less common which take on a brownish look are called **'Monocrystalline'** or **'Single'** **Cells.**

3) The only other one that we have not mentioned is the **'Thin Film' Cell.** The difference to the others is the way the cell is constructed. These do not seem to be quite as efficient as the others but the cost is much lower.

Remember that the electricity generated by solar cells is DC (Direct Current) so there must be, somewhere, a Converter/Inverter to change that DC into AC (Alternating Current).

The other obstacle is night time. Where does the power for my house come from on a night? There are two ways to do this:

1) **Net Metering:** With this the system is arranged so that the consumer gets inverted power DC to AC during the daytime and taps from the electricity supplier during the night time period. Any excess electricity generated by the solar panels during the daytime can be sold back to the electricity supplier. This means that the electricity cost from the supplier will be less.

2) **Batteries:** The other way to obtain power on a night time is to have a set of standby batteries also linked to the inverter to supply the consumer over the night time period. The DC from the battery bank will be fed into the inverter to produce AC.

When thinking about getting solar panels in a particular house there are several points to look out for:

1) **Cost:** How long would it take in saved energy costs to get your money back?

2) **Moving House:** Are you going to move in the not too distant future and how much value does solar energy put onto the property?

QUESTION:
Have you got or would you consider getting solar panels in your house and have you worked out an energy saving time period?

ANSWER:

Fuel Cells

What is the difference between a fuel cell and a battery? They both have electrodes with a potential difference and capable of supplying an EMF to a load. Some fuel cells work at fairly high temperatures of 150–1100 degrees centigrade.

See diagrams and explanations below:

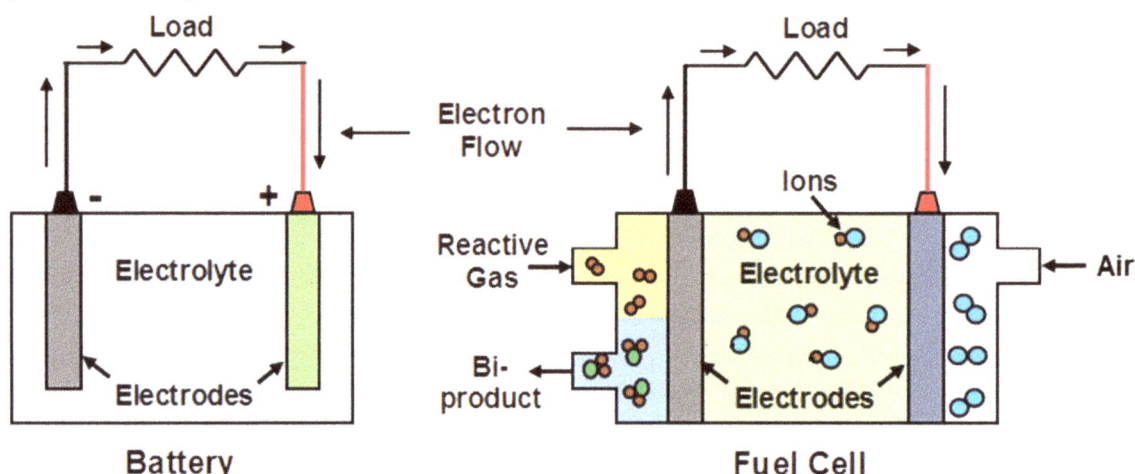

Battery:

Above left, a battery is a self-contained unit consisting of a container with two electrodes and an electrolyte that through a chemical reaction produces a DC voltage. Batteries can be primary or secondary. Primary batteries have a finite life, but when a secondary battery is flat we can connect a battery charger and reverse that reaction and charge the battery back up to what it was in the beginning.

Fuel Cell:

Above right, a fuel cell can do the same job as a battery but is slightly different in the way that it does it. The fuel cell, similar to a battery, has a container, electrodes, electrolyte, and produces a DC voltage through electrochemical reaction, but to achieve the reaction it draws an external energy source into it such as hydrogen and expels a bi-product such as water.

Air is also drawn into the cathode. A catalyst is also required as an extra to assist the reaction to separate the gas ions into protons and electrons. Protons will then pass through the electrolyte. The fuel is usually Hydrogen or gases rich in Hydrogen i.e.: Natural Gas, Methanol, Propane etc.

Air Batteries:

We could in actual fact call an Air Battery a Fuel Cell because to make an air battery work a gas has to be taken in from the outside i.e.: air in this case.

Charging:

As mentioned above secondary batteries require charging when they are flat, but fuel cells do not. So long as the fuel supply, such as Hydrogen and air (Oxygen), is available for the cathode the cell output will continue. So a fuel cell may be classed as a primary battery.

The only by-product produced by the fuel cell is mainly water as a result of the gas reaction. One of the main stumbling blocks for fuel cells in vehicles is the stations available to recharge with fuel gas.

Fuel Cell Types:

As mentioned Fuel Cells are very different from batteries as they require a fuel, usually gas such as Hydrogen, from the outside to make them work. A battery is a self-contained chemical reaction.

All air batteries can be called a type of 'Fuel Cell' as they take air into the cathode to make them work so they use the Oxygen from outside.

1) Alkaline Fuel Cell (AFC)

2) Direct Borohydride Fuel Cell

3) Direct Carbon Fuel Cell (Very High temperature when working)

4) Direct Ethanol Fuel Cell

5) Direct Formic Acid Fuel Cell

6) Direct Methanol Fuel Cell

7) Electro Galvanic Fuel Cell

8) Enzymatic Biofuel Cell

9) Hydrogen Fuel Cells (Covers many listed!)

10) Magnesium Air Fuel Cell

11) Metal Hydride Fuel Cell

12) Microbial Fuel Cell

13) Molten Carbonate Fuel Cell (MCFC) (Very High temperature when working)

14) Phosphoric Acid Fuel Cell (PAFC) (High temperature when working)

15) Planar Solid Oxide Fuel Cell (High temperature when working)

16) Polymer Electrolyte Membrane Fuel Cell

17) Proton Exchange Membrane Fuel Cell (PEM)

18) Protonic Ceramic Fuel Cell (PCFC) (Very High temperature when working)

19) Reformed Methanol Fuel Cell (RMFC)

20) Reversible Fuel Cell (RFC)

21) RFC Redox Fuel Cell

22) Solid Acid Fuel Cell (SAFC) (High temperature when working)

23) Solid Oxide Fuel Cell (SOFC)

24) Solid Polymer Fuel Cell (SPFC)

25) Sulphuric Acid Fuel Cell

26) Tubular Solid Oxide fuel Cell (TSOFC) (Very High temperature when working)

27) Up-flow Microbial Fuel Cell

28) Zinc-Air Battery (Type of fuel cell)

I doubt that there are any more Fuel Cells than those on the list above. There may be some undergoing research, for instance, in the motor vehicle world.

Instrument and Testing Section:

In this section we are going to look at different battery tests that can be done to check battery condition. As I am sure you are aware **ANY** instrument only reads **CURRENT**, anything else such as voltage, ohms etc. is a fictional value and worked out for you by the scale on the test instrument. Let me try and prove this.

I am sure that you all can remember the old AVO analogue multi-meter. t was so accurate that it had to be laid flat and the scale had a mirror at the back where you had to get the pointer level with the reflection in the mirror. The AVO had two sections, AC and DC, and each section had a circular switch. You could switch to DC Amps and Volts or AC Amps and Volts or Ωs. This AVO instrument was moving coil for accuracy which is actually based on a DC motor with a finite movement.

Let us concentrate on DC Amps and Volts. To measure volts I just put across the battery and the scale will give me the Potential Difference of the battery in volts. With amps I put the instrument in series with a circuit and read the EMF amperage. I read these two totally different values **WITH THE SAME INSTRUMENT.**

So if I switch to DC voltage and measure the battery potential difference across my battery, why then if I switched to DC amperage and put the instrument across the battery would the needle violently move across the scale and bend if the safety cut out did not operate? What has changed inside of the instrument? **IMPEDANCE.**

If you opened up the AVO Instrument it would be absolutely full of resistors and diodes. If you switched to **AC VOLTS** then the power would go through diodes and resistors so the instrument would be actually reading **DC AMPS** and the scale would change the reading to volts.

The AC amperage setting would have a different set of resistors in circuit inside of the instrument to DC voltage. Switching to DC amps and putting across the battery would cause the instrument cut out to work because you would actually be measuring the short circuit current which could be huge. Switch to DC volts and the current would be much less and the scale of the instrument would change the reading to voltage for you.

The ohms section of the instrument is slightly different as it uses the internal battery instead of external power, but in the end reads **CURRENT** and converts to **OHMs** on the scale.

IT WOULD BE EXTREMELY DANGEROUS TO SWITCH TO OHMS AND PUT THE LEADS ACROSS A BATTERY AS THE TWO VOLTAGES WOULD NOT BE COMPATIBLE AND THE INSTRUMENT COULD CUT OUT OR EXPLODE.

This section will deal with:

1) Carbon Pile Battery Testing

2) Battery Internal Resistance Testing

3) Battery Conductance Testing

4) Battery Testing with a Multi-meter

5) Thermal Image Testing on a Battery Bank

6) UPS Amp hour Testing of a Battery Bank

7) UPS String Voltage and Current

8) UPS Ripple Voltage and Current

Battery Carbon Pile Voltage Testing:

What we are looking at here is how much voltage and current a battery gives out under a specific load condition. If you were to put a multi-meter onto DC volts and put it across the battery you may get the open circuit voltage of that battery, but your meter would have impedance so you would not get the data with the battery under load conditions. All you would be testing here is that the battery is giving out a voltage so is charged to some extent. To tell if a battery is on its way out we have to put some serious load onto it and test the voltage and current.

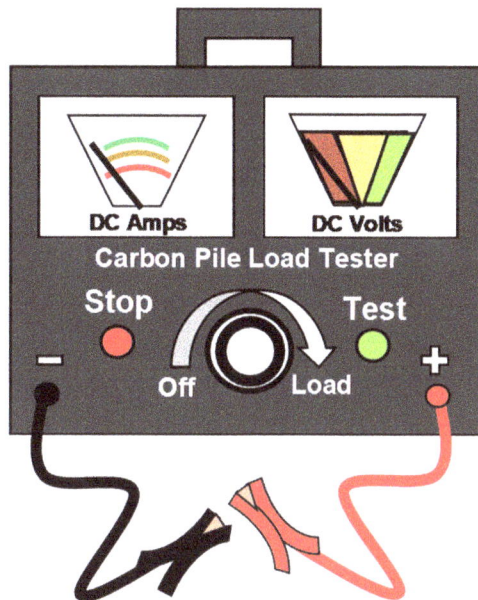

First thing is to ensure that we start the test with a fully charged (at least 80% of full charge) battery and determine whether it has been in service or sitting on a store shelf, and also ensure that the electrolyte is topped up with water.

The tester in the diagram above consists of a large resistor and a load control, which is really a variable resistance or rheostat. It can be adjusted to suit whatever battery is under test. So the positive and negative leads are connected to the battery and the load adjusted for a particular type of battery. The 'Test' button is pressed and a timer puts the large resistor in circuit for a set time and the voltage drop off measured. The test can be halted at any time by pressing the 'Stop' button. Basically we have a stack of Carbon disks under pressure which can be adjusted using the load dial.

Not all testers have an operation procedure the same as I have just described, or it may not look like the diagram at the top, but they all achieve the same objective of measuring the battery readings under load conditions. The voltmeter will have red and green sections and the battery reading wants to be in the green section indicating that the battery is ok.

Ensure that the variable resistor dial is in the 'OFF' position on the test instrument to ensure that there are no sparks when the leads are connected. Connect the tester to the battery, taking care to connect the red lead to the positive terminal and the black lead to the negative terminal. At this point the voltmeter will read 'State of Charge'.

Inspect the battery and locate the Amp hour (Ah) rating. This may be called the 'Cold Cranking Amps' (CCA). Rotate the variable resistor dial to the setting for this battery which will be 50% of the battery Amp hours (Ah). Push the test button to start the test and after 15 seconds (manufacturer's recommended time) check the voltmeter reading. If the reading is green the battery is ok, but if the reading is in the amber or red then check the manufacturer's data as the battery may require replacing.

Safety: Just remember that with quite a lot of batteries Hydrogen (H_2) is given off so be careful in the way of sparking and do not have the test set too close to the battery. When carrying out any battery testing always wear the correct safety PPE especially eye protection and gloves.

Battery Internal Resistance Tester:

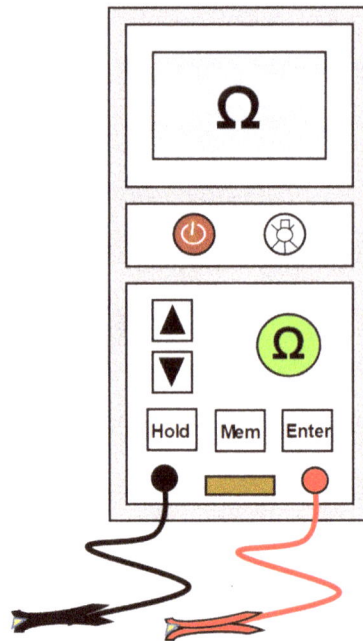

Battery Internal Resistance Testing can be done fairly quickly and instantly indicates the battery internal health. The tester measures the voltage and the resistance of the connections of the internal cells. The testers themselves differ in design and the way they operate, but all achieve the same objective. High resistance readings may indicate that the battery terminal voltage may drop when the battery is under load or the internal connections will heat up.

Just to describe the tester above, first of all it is connected to the battery ensuring that the red lead is connected to the positive and the black lead to the negative. The red on/off button is pressed to switch on and there is a button to light up the screen if required. The green button is pressed to start the test and a reading in ohms comes up on the screen.

The hold button will hold the reading if required. The memory button is pressed and the enter button will allow the reading into the tester's memory should several batteries have to be tested. At the bottom of the instrument there is a facility to connect the tester to a PC to enable comparison to past results and maybe a dynamic graph to be plotted.

QUESTION:

Have you carried out this Resistance Test or a Carbon Pile Test? If so why?

ANSWER:

Battery Conductance Testing:

This test is very similar to the internal battery resistance test and similarly only takes a very short time to complete. Usually completed on lead acid batteries such as the one that powers the electrics in your car.

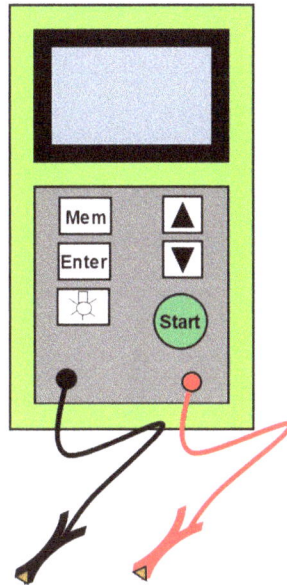

What we are trying to achieve here is a measurement of plate surface availability, which will determine how much power the battery can supply (Cold Cranking Amps CCA).

All through the battery descriptions I have talked about sulphation which is about white crystals forming on the battery internal plates, which could be caused by letting the electrolyte level drop for long periods of time. This sulphation cuts down the efficiency of the internal plate to do its job.

Firstly before the test is started just check over the battery for damage in the way of cracks or bulging, surface corrosion or dirt on the top.

The tester above comes in many shapes and sizes and some differ in the way the test is completed. Connect the tester to the battery ensuring that the red lead is connected to the positive and the black lead to the negative and the clips have a firm connection to the battery terminals. The surface charge should be removed from the battery by putting it under load for a few seconds.

Connecting the instrument to the battery will produce some information on the screen as to the voltage of the battery which may be 12.03–12.04 volts on a car lead acid battery. Push the start button and the test will commence. The next step is to enter the battery information that is requested on the screen by using the up and down keys and 'Enter'. The scale will ask for information such as: what type of battery we are testing (Normal or AGM), the post configuration (Top or Side) and what we want to test for.

Select Cold Cranking Amps and set the scale from the battery data which should be available. Press 'Enter' and the test will commence. After a few seconds more information will appear on the screen such as measured 'Cold Cranking Amps' and 'Voltage' followed by the next screen which will be the 'State of Health' of the battery.

We can predict the end of life of the battery. As you can see there is a memory button so that results can be stored and later downloaded to a PC. We can then see the 'trend' and if the particular battery is getting better or worse.

'Open Circuit'/Surface Charge testing with a Multi-meter:

A multi-meter is an extremely handy instrument because of its multi-settings. This instrument however has a high input impedance and small current so if you were to put it across say a 12-volt Lead Acid battery, of the type used in a vehicle, you would actually get the battery open circuit voltage/surface charge which would be 12 volts+. Because of the impedance/low current you may even get this reading if the battery was nearly flat. You would only find a battery fault with this instrument if there was to be some catastrophic failure with the battery.

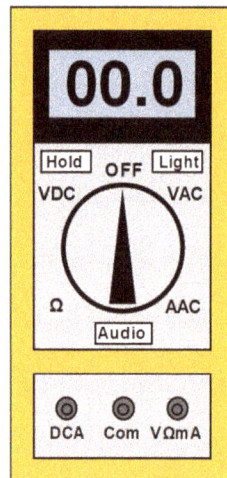

Let us say that we are going to test a 12-volt Lead Acid battery of the type used in a vehicle. To test the battery using this instrument, plug the test leads into 'common' (black lead) and 'volts' (red lead) at the bottom. Switch the changeover switch to DC 20 volts and put the clips/leads onto the battery terminals. The reading would be the battery DC volts at that time which would be around 12 volts+. This would be called the 'Open Circuit' or 'Surface Charge'.

To obtain a true voltage reading the battery must be put onto some sort of substantial load with the multi-meter connected across the terminals in parallel. So if this was a car then switching all of the lights on including the headlights or the turning of the starter motor and starting the engine might be sufficient.

Before carrying out testing using this instrument, many technicians test the instrument first by switching the changeover switch to 'Ω' and putting the lead clips together, obtaining a reading of around 0.4Ω–0.8Ω which would be the resistance of the leads. Sometimes they forget to switch the changeover switch to DC volts before connecting across the battery.

THIS INSTRUMENT MUST NOT BE CONNECTED TO THE BATTERY WITH THE SETTING ON OHMS OTHERWISE THE INSTRUMENT MAY SUFFER CATASTROPHIC DAMAGE OR EXPLODE!

QUESTION:

Have you carried out a Conductance Test or Open Circuit Test? If so why?

ANSWER:

Thermal Imaging:

Thermal Imaging can save a battery system from suffering a catastrophic breakdown due to heat caused by a high resistance connection or a faulty battery. The instrument I have shown below is a standard Instrument, but Atex Thermal Imaging Devices can be obtained and may be required in, say, a battery room where there may be **Hydrogen Gas.**

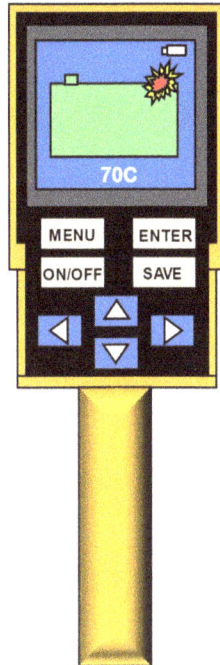

The battery bank really needs to be under load conditions to complete this test. Switch the UPS system to by-pass and put the batteries under load. Wait around 15 minutes to see if things start heating up.

The instrument is very easy to use, simply push the **'ON/OFF'** button and a picture will appear on the small screen. Start at the beginning of a line of batteries and very slowly walk along the line pointing the instrument at the batteries.

The picture remains green for normal temperature, but will turn red & yellow around any points of heat. This could be anything from the whole battery going faulty to a high resistance connection.

The 'MENU' on the device will give several options so you will push 'ENTER' to select the one that is appropriate for the test you are carrying out. The information can be saved and downloaded to a PC at the end of the inspection. You may end up with a trend where something is slowly getting worse.

As mentioned above it is possible to obtain an Atex Thermal Imaging Device which may be required in a battery room with **Hydrogen Gas.**

What can cause heat sum up:

1) Loose connections of battery linkage wiring.

2) Lugs not properly crimped on battery linkage wiring.

3) Batteries that are developing failure.

4) Overcharging.

Battery Bank Amp Hour (Ah) Test:

Let us select for an example a UPS System feeding the Emergency Lighting System on plant. In this test what we want to check is actually how long the Emergency Lighting will stay on before the inverter trips out on low voltage. This will give us the battery bank Amp hours. This is called **'Battery Autonomy'**.

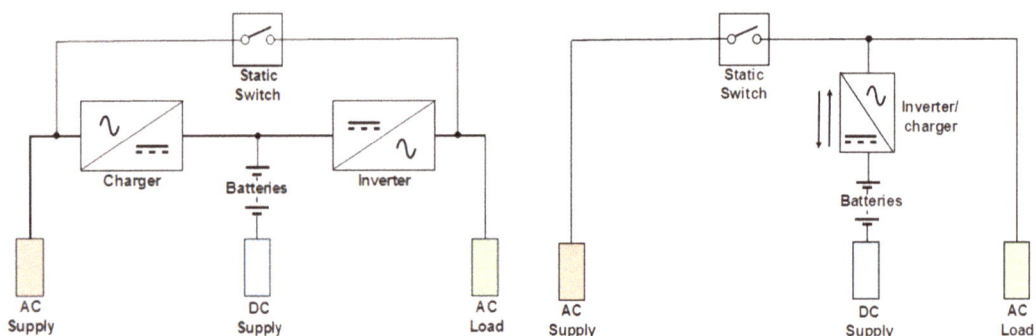

Operate the Static Switch so that the plant emergency lighting is fed via the inverter as they would be in a power outage. What we don't want is the plant emergency lighting tripping early and putting the whole plant into complete darkness!

The danger here is, and this is one of the values we are looking for, that some batteries do not like going completely flat so the inverter must trip before the voltage of the batteries drops below the minimum manufacturer's recommendation so that damage is not done to the batteries. This is one of the test checks to be done and this would be called the **'Cut off Voltage'**.

This test also ensures that all of the designated emergency lighting fittings are ok and working. It is tempting to do this test on a night so that the fittings can be easily checked. The problem here of course is fairly obvious. If when it gets dark you operate the static switch and the inverter kicks in to supply the emergency lighting from the battery bank and the inverter trips, if at that moment there was to be a power outage the plant would be left with no main or emergency lighting and would be in complete darkness until the power returned.

So do this test first thing in the morning to give the charging system time to charge up the battery bank before nightfall.

So we are looking for:

1) How long the lights are on under inverter/battery power.

2) Are all of the fittings on site in good maintained order? I.e.: no lamps out.

3) Check the manufacturer's minimum voltage of the batteries in the bank.

4) What voltage does the inverter trip operate?

5) This would be an ideal time to carry out thermal imaging on the batteries and cabling.

6) What current is the system taking?

7) Do the emergency lighting site isolation switches operate ok?

At the same time the UPS can be checked for:

1) 1 – UPS string/battery bank charge/discharge voltage and current.

2) 2 – UPS ripple/unwelcome voltage and current caused by extra oscillating AC load from the charger.

Some UPS Systems actually complete a battery check daily by putting a load onto their battery bank and alarming if there is a problem. This only applies to the full bank rather than individual batteries.

Calculations – Resistance, Voltage & Amp Hours:

Resistance Calculations:

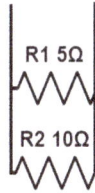

R1 5Ω

R2 10Ω

$$R\ Total\ =\ \frac{R1 \times R2}{R1 + R2}\ =\ \frac{5\Omega \times 10\Omega}{5\Omega + 10\Omega}\ =\ \frac{50\Omega}{15\Omega}\ =\ 3.33\Omega$$

Resistors in Parallel

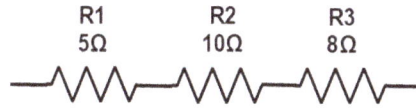

R1 5Ω   R2 10Ω   R3 8Ω

Total Resistance = 5Ω + 10Ω + 8Ω = 23Ω

Resistors in Series

To calculate total resistances in series and parallel is as per formulas above. In series just add them together,r in parallel just a bit more complex: R1 x R2 over R1 + R2.

Battery Voltage Calculations:

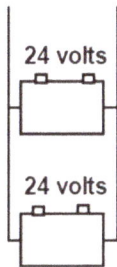

24 volts

24 volts

Voltage = 24 volts

Batteries in Parallel

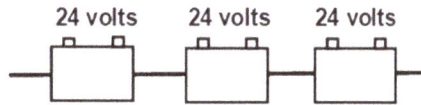

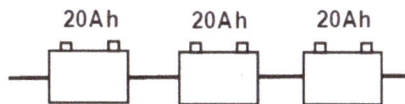

24 volts   24 volts   24 volts

Voltage = 24 + 24 + 24 = 72 volts

Batteries in Series

Batteries in parallel add up to the same voltage as 1 battery, but batteries in series are added up.

Battery Amp Hour Calculations:

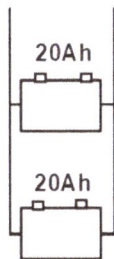

20Ah

20Ah

Amp hours = 20 + 20 = 40Ah

Batteries in Parallel

20Ah   20Ah   20Ah

Amp hours = 20Ah

Batteries in Series

Batteries in parallel add up the Amp hours (Ah), but with batteries in series the voltages are added together. So in a rack the batteries would be connected in both series and parallel. Series to obtain the voltage and parallel for the Amp hours.

Several Examples of Battery Terminology:

Absorption Charge

If a battery is completely flat a long charge may be required to bring it up to full charge. When 80% of the battery charge capacity is reached at the rate of charge, it is called the **Absorption Charge**. The current should taper off and, depending upon the battery and how flat it is, the charge to this level could take several hours.

AGM (Absorbed Glass Matt)

The mat is a fibreglass material a bit like a sponge for electrolyte and is between the plates keeping the electrolyte where it is required regardless of the position of the battery.

Ah (Amp hour- Ah)

Battery sizes are measured in **Amp Hours (Ah)** which is the amount of energy in the battery which will allow 1 Amp to flow for 1hour.

Anions

Negative atoms/ions that have gained electrons so more electrons than protons.

Anode (Battery Negative-)

Battery negative and reservoir for anions.

Aqueous Electrolyte

Water-based electrolyte.

Battery Bank

A set of batteries connected in series and parallel supplying a UPS System.

BESS (Battery Energy Storage System)

Could be described as the **Battery Bank** which feeds a **UPS System.**

BMS (Battery Management System)

A system that looks after a rechargeable battery set up by monitoring and keeping the battery(s) within its safe working mode.

Boost Charge

Short duration charge of high current to charge a battery up very quickly.

C Rate or Charge Rate

When we look at the '**Charging Rate**' this is called the '**C' Rate** which is the time it takes the battery to charge, i.e.: 2C is 30 minutes, 1C is 1 hour, 0.5C is 2 hours etc. How do we find this 'C' Rate? Well on small secondary batteries the 'C' rate should be either on the side of the battery or on the datasheet. Not all types of battery will be the same. On larger batteries the '**C' Rate** is worked out using the Amp hour (Ah) rating of the battery. So to work 'C' out:- t (time) = 1 (hour) divided by the 'C' rate. The answer will be in hours.

Carbon Pile Voltage

What we are looking at here is how much voltage and current a battery gives out under a specific load condition.

Catalytic Cap

Designed to recycle gases that are released through **'Gassing'**.

Cathode (Battery Positive +)

Battery positive and reservoir for **Cations**.

Cations

Positive atoms/ions that have lost electrons so fewer electrons than protons.

CCA (**C**old **C**ranking **A**mps)

Cold Cranking Amps is how many amps a battery can deliver for 30 seconds in a temperature of 18 degrees centigrade.

CCV (**C**losed **C**ircuit **V**oltage)

Closed Circuit Voltage is the voltage of the battery under load conditions.

Cell

Batteries are made up of many cells, which could be called energy producing modules, connected together inside of the battery.

Charge Acceptance Current

The amount of charge that a battery can accept in normal conditions in a set time.

Charging Current

The '**Charging Current**' of a battery should be 10% of the amp hour (Ah) rating so a 150Ah battery charging current would be 150 x 10(%) over 100(%) which would equal 15 amps.

Charge Efficiency

This is to do with how much energy was removed from the battery when it was working (discharging) compared with how much energy is used to bring the battery to full charge. Sometimes referred to as '**Charge Acceptance'**.

Conductance Tests of Batteries

What we are trying to achieve here is a measurement of plate surface availability, which will determine how much power the battery can supply (Cold Cranking Amps CCA).

Constant Current Charge

A Constant Current Charger will deliver, once set with a variable resistance in series with the battery, constantly and without variation regardless of the battery being fully charged or not.

Constant Voltage Charge

A Constant Voltage Charger is usually the common car battery charger which just has a mains feed through a rectifier and via a variable resistance in parallel, feeds straight onto the battery and will not vary until switched off.

Cycle Life

When a battery fully discharges and then is charged up again, it is called **1 Cycle.** The manufacturers will determine how many cycles the battery will do to the end of its life.

Cut Off Voltage

UPS batteries for instance will have a 'Cut Off Voltage' which is the minimum voltage they can go down to before they would be completely flat and suffer possible damage.

CID (Current Interrupt Device)

Similar to a fuse inside the battery, especially on Lithium batteries.

Deionised Water

This water is completely different from distilled water. Often referred to as demineralised water. (Used in large industrial boilers.)The water is passed through electrically charged cation & anion resin beds to remove its ions which tap water would be full of. As it passes through the resin filter, impurities are attracted out of the water as they are attracted to the opposite charges.

Distilled Water

Distilled water is sometimes called 'Pure' water and does not contain any of the impurities found in tap water i.e.: Sodium (Na), Calcium (Ca2), Magnesium (Mg2) so there are no minerals to cause 'sulphation' or ions to cause rogue electron flow. So this water is ideal for wet/flooded batteries. Distilled water is made by a system as you might guess called 'distillation' which is a bit like catching the steam coming out of the spout of a kettle. **Ideal for topping up batteries.** This water is good for you to drink, possibly chilled, but may taste bland compared to spring water or tap water.

DoD (Depth of Discharge)

Depth of Discharge is how much in % the battery has been discharged from its relative full charge.

Dry Charged

Sometimes the manufacturers 'Dry Charge' a flooded battery with no electrolyte.

EFB (Enhanced Flooded Battery)

An **Enhanced Flooded Battery** is an up-rated version of a wet battery (Liquid Electrolyte).

Electro Chemical Potential

The ability of a metallic material anode to lose electrons. Lithium of all the elements has a much higher tendency to lose electrons than nickel.

Electrolyte

Electrolyte is a liquid, gel or solid substance that is a vital part of a battery chemical reaction.

Electrolytic Cells

Electrolytic Cells are the units that can be used in **Electrolysis** (a non-spontaneous Redox Reaction) and require an external source of energy or in other words require a DC power supply.

Element

An **Element** in battery terms refers to the anode, separator and cathode together as a unit.

Energy Density

This is the amount of **Energy Stored** compared with the **Volume**.

EOL (End of Battery Life)

The battery can deliver an output of only 80% maximum or less of full efficiency.

Equalisation Charge

Used on, say, a UPS battery bank after a number of cycles or a time period to bring all of the batteries up to the same level of 100% charged.

ESR (Equivalent Series Resistor)

This is the Resistance of the internals of the battery which limits the amount of current.

Float Charge

A **Float Charge** is carried out by a **Smart Charger** which only charges the batteries when they require charging as against a **Trickle Charge** which is on all of the time.

Flooded Battery

Another name for a **Wet Battery** with liquid electrolyte.

Flow Batteries

These batteries have pumps to pump the electrolyte around. If the pumps are not working the battery is dormant.

Formation

Formation is charging the battery for the very first time. (Manufacturing)

Fuel Cell

A fuel cell can do the same job as a battery but is slightly different in the way that it does it. The fuel cell, similar to a battery, has a container, electrodes and electrolyte and produces a DC voltage through electrochemical reaction, but to achieve the reaction it draws an external energy source into it such as Hydrogen and expels a bi-product such as water.

Galvanic Batteries/Voltaic Battery

This **Galvanic Battery** is what the book is about. A Galvanic or **Voltaic Battery** is a device which turns energy from a chemical reaction into DC electrical energy.

Gassing

Caused, for example, by overcharging. The cells release Hydrogen & Oxygen gas.

Grid

A framework which supports the plate active material.

Hydrometer

A device for measuring the electrolyte SG.

Hysteresis

'**Hysteresis**' can occur due to the chemical reaction lagging the charge voltage.

Inductive Charge

Uses **Electromagnetic Induction** to charge portable equipment such as smart phones, instead of physical leads.

Internal Resistance of a Battery

Battery Internal Resistance Testing can be done fairly quickly and instantly indicates the battery internal health. The tester measures the voltage and the resistance of the connections of the internal cells.

Inverter

A device for turning DC (Direct Current) Battery Power into AC (Alternating Current) by creating a Sine Wave AC Power System.

Ions

Particles called **ions,** some of which are called cations which, unlike natural atoms, have more protons than electrons so they are positive, and some of which are called anions which have more electrons than protons so they are negative, gather on each electrode.

Jelly Roll

This is the way the manufacturers make the **Cylinder Battery**. They put all the ingredients together along with separators, electrodes and electrolyte in a flat layered form and then role it into a tube like seaside rock and this action is called a **'Jelly Roll'**. The objective is to ensure that the electrodes are more in contact with the electrolyte.

MCA (Marine Cranking Amps)

Similar to Cold Cranking Amps (CCA) but carried out at a higher temperature of around 32 degrees instead of 0.

Memory Effect

A minority of older batteries have a Memory Effect which means that if you were to keep charging them before their charge was depleted a 'Memory' would build itself into the battery where it would only accept a percentage of Full Charge. Nickel Cadmium springs to mind.

MF (Maintenance Free)

A Maintenance Free or sealed battery does not require topping up with distilled water.

Minimum Charge

Is there a minimum charge voltage that I have to put into a battery so that it accepts the charge? Most definitely and that voltage is over 2.15 volts/cell. Anything less will not charge.

Mossing

Positive Material in the Electrolyte which could cause an internal short circuit.

Nano Battery

Very small battery used in Nano Technology.

Nominal Voltage

This is the voltage of the battery when it is half way between being fully charged and very low charge.

OCV (Open Circuit Voltage)

Open Circuit Voltage is, say, testing the battery with a multi-meter without any load. This would be higher than the battery under load.

Operating Temperature

Batteries have a temperature in which they operate which will be determined by the manufacturers having carried out numerous tests. If the user temperature was outside these parameters then the battery life, charge and efficiency would be affected.

Over Charge

Overcharging a battery is when the charge is left on for too long. This is easier to achieve on, say, constant **Boost Charging.**

Over Discharge

It is possible to over discharge a battery and let it go completely flat below manufacturer's tolerance. Damage may occur to the battery if this happens. UPS systems have a 'Cut Off' Voltage before this happens.

PCM (Protective Circuit Module)

A Smart Electronic System that prevents batteries over-charging and over-discharging.

Power Density

This is the amount of Power Output compared with the Volume. Not as widely used as Energy Density which is slightly different.

Primary Battery

A Primary battery cannot be recharged.

pH (Power of Hydrogen)

Expresses acidity or alkalinity.

PTC (Positive Temperature Coefficient)

This is like a thermistor/Thermocouple Switch inside of the battery which cuts out if the temperature reaches a certain point and then resets as things cool down.

Pulsed Charge

A '**Pulsed Charge**' requires a special battery charger to give bursts of charge to the battery, say, a few milliseconds apart. The idea is that during charging what we are actually doing is causing the chemical reaction that happened during the battery discharge to be reversed. Well the charger giving the battery pulses of charge allows the reverse chemical reaction going on inside of the battery to keep up with the charge. If this is not done a phenomena called '**Hysteresis**' can occur due to the chemical reaction lagging the charge voltage.

PV (Photovoltaic)

This is the conversion of light into electric energy. Solar Cells are the classic example.

RC (Reserve Capacity)

Reserve Capacity is how long the battery can continue to supply should charging systems fail. A good example would be a car if the alternator charging failed.

Redox (Reduction Oxidation)

This is a process where one chemical's atoms lose electrons and others gain electrons. Flow batteries are a good example.

Safety Gas Venting System

Some batteries have this system which is there to allow gases inside of the battery produced during charging to vent out. The actual container is usually very strong so that these gases do not cause it to 'bulge' or rupture, but in extreme cases batteries have been known to explode even with a strong case.

Secondary Battery

A Secondary battery is one that can be recharged.

Sediment

Material in the bottom of the battery which has fallen from the plates.

Self-Discharge

Batteries, as we have discussed, are chemical reactions resulting in DC electricity. With most batteries this chemical reaction ceases when the load is removed, but a minority of batteries keep reacting even with no load. This is also to do with **Shelf Life.**

Separator

A Separator is a physical shield inside of the battery to stop the anode from touching the cathode.

Shelf Life

As above, this is how long the battery can be stored before it starts to degrade, meaning it will not last as long when purchased. A minority of batteries self-discharge so their **Shelf Life** will be short.

SLA (Sealed Lead Acid)

A **Maintenance Free** or sealed battery that does not require topping up with distilled water.

SLI (Starting, Lighting, Ignition)

An SLI battery is a name for a vehicle battery.

Smart Battery

Battery with internal circuitry designed to communicate information.

Smart/Intelligent Charger

Battery charger with **Management System** for charging.

Specific Gravity

This is actually a measure of the weight of acid or alkaline when mixed with a weight of water.

Splash Barrel

This is part of a **Battery Venting System** to try and keep electrolyte out of the venting system should the battery get shaken about when the battery is flat. A car battery springs to mind.

SoC (State of Charge)

This is a measure of how much battery capacity there is left at any one time.

SoH (State of Health)

This is a battery's ability to deliver its energy at any one moment in time.

Stratification

Stratification occurs when, say, a wet battery is not used for some time and the acid separates from the water and sinks to the bottom of the battery.

Sulphation/Sulfation

This occurs if the electrolyte level is allowed to drop and Sulphur crystals form on the plates. This can also happen in Sulphuric electrolyte batteries where the Sulphur separates in the electrolyte and forms on the plates.

Thermal Runaway

A thermal runaway can be termed as an **Uncontrollable Exothermic Reaction**. Let us say that the battery becomes internally damaged to, for example, short circuit stage. The amount of heat generated inside of the battery would be enormous and unstoppable. This is why a separator may be used.

THD (Total Harmonic Distortion) (UPS Systems)

This is the amount of distortion in **Sine Waves** against their ideal formation. Usually associated with **UPS Systems** where **inverters** may be used to create the sine waves.

Traction Battery

A very heavy duty battery for large loads.

Trickle Charge

Trickle Charge is a low constant charge that is put onto a battery from say a **Linear Charger** that does not vary as against a **Float Charge**.

UPS (Uninterruptible Power Supply)

A system which supplies emergency power in the event of a main power outage usually with the assistance of an inverter.

Voltaic Battery/Galvanic Battery

Converts chemical energy to DC electricity.

VRLA (Valve Regulated Lead Acid)

These are more commonly known as **Sealed** or **Maintenance Free** batteries.

Wet Charged

This is the common way to charge a flooded battery with the electrolyte in as against a dry charge where there is no electrolyte.

Wh (Watt hours)

A **Watt hour** is a measurement of power over a set time i.e.: 1 hour. So 1 watt for 10 hours would equal 10 Watt hours. Electricity meters measure Kilowatt hours (KWh) so that the electricity company knows how much electricity has been used.

Battery Room PPE:

The Battery Room:

The room should be well ventilated with **Atex Certified Fans and Lighting** as per company policy. Remember the gas from many batteries is Hydrogen and/or Oxygen. Nothing should be stored in a battery room except battery tools, filling units, funnels and locker for PPE etc.

Battery Racks:

Battery Racks should be clearly marked as to system so that the wrong batteries cannot be chosen. The type of battery i.e.: acid or alkaline should be marked so that equipment used on acid batteries is not used on alkaline or otherwise acid could neutralise alkaline or vice versa.

Battery Chargers/Inverters/Converters:

Battery chargers and inverters etc. should be switched off whilst the batteries are under maintenance. Process and other concerned departments should be made aware that this is happening as various alarms may come up immediately or during the work.

Fire Fighting Equipment:

Correct Fire Fighting Equipment should be readily available in the room and clearly marked. This obviously may eliminate water or foam extinguishers. People should be converse in the use of fire-fighting equipment and only tackle small fires otherwise evacuate. **Usually CO2 or Powder**.

Short Circuits/Fire:

Since batteries cannot be isolated there is always the risk of accidental short circuits, explosions and fire. Personnel should be made aware of this in the above training. Short circuits can result in very high energy arcing.

Gas Detectors:

It may not be a bad idea to have personnel carry a **Gas Detector** which will indicate the level of gas in the room i.e.: Hydrogen and Oxygen content in air. Remember that if the **Risk Assessment** states that Oxygen is given off then grease may have to be limited as pure Oxygen and grease are explosive.

Warning Signs:

There should be large, clear signage around the room indicating PPE to be worn when maintenance is being carried out on batteries. I think it goes without saying that no smoking should be carried out in the room with signage indicating this. Copies of regulations and resuscitation procedure should also be displayed.

Eye Wash Stations:

Several Eye Wash Stations should be available in any battery room with the **SEALED** bottled saline solution in date and bottles should be clean and clearly signed. If a bottle seal is broken, a replacement must be installed. Ensure that instructions are read and followed in an emergency. Sometimes a water spray eye wash station is available which is much better.

Spillages:

There should be a container full of material for spreading onto a spillage just in case. Spillages should be dealt with immediately. Overfilling the battery can cause corrosive electrolyte to spill out.

Training:

Personnel should be given **Competency Training** on battery maintenance. This can be achieved with the use of safety films and an expert talk. Emergency procedure and resuscitation training should also be carried out. Training to include regulations appertaining to battery maintenance.

Battery Contents:

Maintaining flooded or wet batteries can be hazardous. Remember what the batteries contain, acid or alkaline, both of which can burn skin and be **VERY** hazardous to eyes. Batteries have also been known to explode so it is important that the correct PPE is worn. As above, tools/instruments which come into contact with acid should not be used on alkaline or vice versa. Any electrolyte contact with skin should be washed off thoroughly with water. Spillage should be dealt with immediately.

Personal Equipment/Jewellery:

It may be wise to put insulation tape over any rings even though gloves will be worn, and have no instances where neck worn jewellery can dangle outside of clothing. Watches should be removed and put in a safe location. Under no circumstances must standard **Mobile** Phones or Radios be allowed in the room as **Electro-magnetic Radiation** is an ignition risk.

Electric Shock:

I know that I state the obvious that batteries cannot be isolated although individual racks may well be. Remember that the total electrical output from batteries is **DC** and not **AC** so the implications of this should be relayed to the persons carrying out the maintenance.

Food and Drink:

NO food or drink must be allowed into the battery room for fear of contamination. Acid does not directly evaporate such as the water, but **Acid Mist** may be possible when lids are removed.

PPE Location:

PPE should be kept clean in a locker type unit. No PPE such as goggles should be left lying about as dust and dirt will collect on them. PPE such as gloves, aprons etc. should we washed after maintenance is complete in case of contamination.

Tools:

Tools should be insulated and non-sparking as far as reasonably practicable and kept in a storage facility in the battery room. Torches should obviously be Atex safety torches.

Cleaning Battery Tops:

Ensure all caps are tight and use a small brush or a cloth to clean the battery top. **DO NOT USE WATER, EVEN DISTILLED WATER!**

Liquids:

Liquids should be in appropriate containers and clearly marked. Especially acid or alkaline etc.

Risk Assessment:

People usually think that these **Risk Assessments** are tedious to complete and a waste of time, but I can assure you that they are not! A Risk Assessment, of all that we have mentioned up to now, should be carried out. It does not have to be complex and will act as a reminder. See copy on page 104.

Lesson Plan:

A **Lesson Plan** or **Safe System of Work** should be carried out on the type of batteries that are to be maintained. How is the maintenance going to be carried out and using what tools and equipment? Manufacturer's instructions and checklists may be consulted.

PPE:

Ensure all PPE to be used is clean and in good order before and after use.

PPE Safety Glasses:

Safety glasses do not seal around your eyes. All I can say here is **better than nothing!**

PPE Goggles:

There are two distinct types of goggles, **Primary and Secondary. Primary** goggles ARE NOT SUITABLE for battery maintenance as they have holes in the top and bottom to allow air flow through the space inside to stop them steaming up. Acid or alkaline can also go through these holes. **Secondary** goggles have filters in the top to allow the air flow and **THESE ARE SUITABLE**! Goggles provide a full seal around your eyes. Clean after completion.

PPE Face Visor:

Very suitable for protecting face and eyes, on a par with **Secondary Goggles** as far as eye protection is concerned. Clean after completion.

PPE Gloves:

Suitable **Thick Rubber Type Gloves** should be worn that are acid and alkaline resistant. I always recommend thin rubber gloves under the larger ones for universal use unless they are your own. People do have skin problems. Clean after completion.

PPE Rubber Apron:

It is well to wear a **Rubber Apron** over any overalls and no bare arms. If electrolyte was to splash onto overalls maybe you were unaware, then when they are washed you may find them full of small holes. Clean/wash rubber apron after completion.

PPE Mask:

Acid Mist may be exerted when a battery cap is removed. Try to avoid close contact with face or wear a mask as per **Company Policy** or **Manufacturer's advice.**

Equipment – Filling Unit:

It is possible to obtain a filling unit where you charge up the unit with a built-in hand pump, remove the battery cap and put the nozzle into the hole up to the fill level which is set on the gun. You then squeeze the trigger and distilled water is ejected into the battery and a small alarm sounds when the correct level is reached. Clean after completion.

Equipment – Plastic Container:

Holds the distilled water for topping up the old <u>fashioned</u> way. Clean after completion.

Equipment – Plastic Funnel:

This is used to top up the battery without spilling distilled water onto the top. Clean after completion.

State of the Art Battery Maintenance Equipment:

I thought I would show some of the newest or little-known technology which can make battery maintenance easier. UPS systems require banks of batteries that have several tens of batteries in them. Maintenance can be a huge, messy task that nobody really wants to do, but if we can obtain technology that makes this task easier and quicker, we not only save money but make the things much more interesting for the technician doing the work. You may have heard of, or even be using, the instruments or equipment below, but if you have not then it may be worth giving a thought to obtaining some of it. Remember that battery operated equipment may have to be Atex Certified. Always check with manufacturers and err on the side of caution!

Electrolyte level Sensor:

It is possible to now obtain an **Electrolyte Sensor Valve** which is inserted into one or many batteries of a bank. Some indicate level via an alarm consisting of LEDs which will alert if the electrolyte level is getting low and the batteries require top up. Others use a type of float unit. There are many different types on the market.

Water Deioniser Unit:

You can obtain a **Deionising Unit** which will deionise, say, tap water into pure water. A cartridge is inserted into the unit which can purify many hundreds of gallons of water.

Battery Top-up Unit and Gun:

These **Top-up Units** are plastic containers which can be pressurised. Holding several gallons of water, it has a hose and gun. Insert the nozzle of the gun into the battery and squeeze the trigger and the water will be inserted into the battery and shut off at a pre-set value when the correct level of electrolyte is achieved. This means batteries can be filled very quickly without any danger of being over-filled. Simple units without a cut off can also be obtained.

Digital Hydrometers:

These **Digital Hydrometers** give a direct reading of the battery specific gravity without relying on a float. There are many makes on the market which are very user friendly with easy to read scales. Remember especially this device may have to be Atex Certified.

Battery Analysers:

Pocket size **Battery Analysers** are very handy for just clipping onto the battery terminals and obtaining the status of the battery and remaining life. It would be very handy to know what the battery CCA is and whether it is going to fail in the near future. Easily used with clear displays and safety devices. Many makes and types on the market. Battery operated equipment may have to be Atex Certified.

Ripple Meter:

State of the art Ripple Meters can be obtained to check ripple voltages of both AC and DC from a UPS System. Ripple Voltage is UPS ripple/unwelcome voltage and current caused by extra oscillating AC load from the charger and can cause batteries to prematurely age. It should remain below 5% of battery capacity. Battery operated devices may have to be Atex Certified.

Thermal Image Device:

Handy for detecting heat from batteries or battery connection. Just aim the device at the batteries and connections in the racks and the display will detect heat immediately. Remember battery operated devices may have to be Atex Certified.

Common Questions about Batteries:

1) **Question:** What is a battery?

   Answer: A battery is a device containing two different elements and an electrolyte which, by a chemical reaction between them, produces DC electricity.

2) **Question:** What is battery charging?

   Answer: When the above chemical reaction is over and completely different chemicals from the original have been produced, the battery is flat. In primary batteries that is the end of their life, but with secondary batteries a battery charger can reverse the chemical reaction and change the elements and electrolyte back to what they were originally. The battery would now be charged.

3) **Question:** What is a separator?

   Answer: A separator, usually porous to allow the chemicals to combine, is a shield in the electrolyte to stop the anode touching the cathode if the battery was shaken or knocked.

4) **Question:** What is an anode?

   Answer: An anode is a battery negative and is associated with 'anions'.

5) **Question:** What is a cathode?

   Answer: A cathode is a battery positive and associated with 'cations'.

6) **Question:** What is an electrolyte?

   Answer: An electrolyte is a material, usually alkaline or acid, that when mixed with a 'polar' solution i.e.: distilled water, produces a liquid that will electrically conduct. Electrolytes in some batteries can be liquid, gel or solid.

7) **Question:** What is the difference between 'electrolytic' cells and 'galvanic' batteries?

   Answer: Electrolytic cells that require an external source of supply and are used in electrolysis. Galvanic or voltaic batteries are what this book is about.

8) **Question:** Why are there not more battery cars?

   Answer: Battery cars are becoming more popular. One big stumbling block at the moment is 'speed of charging'. You pull into a petrol station for a fill up of petrol, how long does it take, 10 minutes? You pull in with an electric vehicle, how long does it take to recharge the battery? Well in Israel they have developed a battery that can fully charge in **5 minutes.** The problems faced now, apparently, are the outputs of the chargers. We are getting there!

9) **Question:** Can I put tap water in my battery?

   Answer: Definitely **NOT** and neither should you use bottled water! These waters contain minerals that are ok for humans, but harmful to batteries. Distilled or deionised water must be used.

10) **Question:** What is the difference between a fuel cell and a battery?

    Answer: A battery stores internal energy and by a chemical reaction emits electrons when loaded. A fuel cell takes an energy source in and turns it into electrical energy which can be a gas.

11) **Question:** What is the difference between a capacitor and a battery? They both store energy!

    Answer: A capacitor stores energy from an external source on plates and releases it in one go. A battery uses a chemical reaction to produce electrons when required.

Battery/Battery Room Safety:

I am going to take this page to state very important safety advice to do with batteries.

Button Batteries:

Unfortunately these are sweet sized and can be easily swallowed by a child. Do not take the idea that if swallowed they will go through the system and come out again! If the battery was to lodge in the stomach or colon they are VERY, VERY CORROSIVE and will corrode their way through stomach linings, tracts and arteries in the body. Take the child to A & E IMMEDIATELY!

Battery Acid/Alkaline:

Battery Acid is very hazardous in the way of burns if there is prolonged contact with the skin so always wear the recommended protective gloves when handling. Eyes are another vulnerable area so recommended eye protection must be worn. If there are any respiratory problems, evacuate the area. Ensure you know where all of the emergency eye wash bottles are.

Lithium and Sodium fire risk:

Both Lithium and Sodium are used in batteries and are both combustible if they get damp and can spontaneously combust. Old batteries must be disposed of correctly and not in dustbins or skips.

Short Circuits:

Individual batteries cannot be isolated so there is always the possibility of short circuits when working on them and arcs can be huge because of the large current that can flow. Metal tools should be insulated and personnel competent. Never purposely short circuit a battery for any reason as the battery under worst case scenario can explode regardless of type or size. Remove any jewellery when maintaining batteries such as watches and put insulation tape around any rings.

Hydrogen Gas:

Some batteries can give off Hydrogen gas whilst charging. I think it goes without saying that there must be no smoking or means of ignition anywhere in the vicinity when carrying out maintenance. Lighting in a battery room must be Atex certified to Hydrogen (llC) and circulation fans should be risk assessed to possible Atex certified standard.

Burning batteries on a fire:

This is unlikely to be the case with large flooded batteries but it might be possible to accidentally throw away smaller cylinder batteries such as AAA, AA, C or D with rubbish onto a bonfire. It is almost certain that these batteries will explode and it is not impossible for them to fly out of the fire like a bullet possibly causing injury.

Battery Overcharging:

With older flooded batteries, overcharging causes the risk of heat build-up inside of the battery which can result in the buckling of the plates causing the material to fall off and the battery to explode. An example might be that you put the battery on trickle charge and forget about it. The standard charger will not back off when the battery is fully charged it will just keep charging up to overcharge.

Battery Room Safety:

Always follow all of the safety precautions mentioned earlier when maintaining batteries and ensure that eye wash solutions and fire extinguishers are in date. Ensure that all safety equipment is stored correctly and is clean and washed before use.

Battery Chargers:

Smart Chargers:

Many industrial battery chargers for UPS systems are what are called **'Smart'** battery chargers which monitor the battery's charge condition and put more or less current onto the terminals or cease charging altogether. If the battery is very flat then the charger will put the equivalent of a boost charge onto the terminals until the battery 'state' comes up to around 86% and will then switch to float or trickle charge accordingly. These chargers are very expensive and fairly large.

Induction Chargers:

With these chargers the device is put **NEAR** the charger, it does not have to be hard wired to the charger at all. An example of this might be a rechargeable electric toothbrush.

Constant Current Charger:

Below I have drawn a very simple diagram of a **'Constant Current'** Battery Charger just to give you an idea of what is actually happening with the different charges.

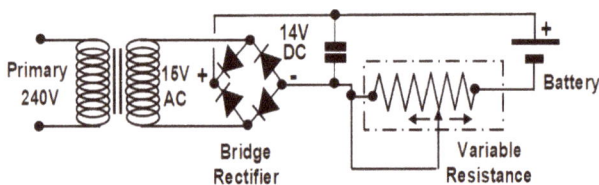

The Voltage comes into the charger via a transformer which is 240 volt AC primary to around 15 volt AC secondary allowing for a slight voltage drop across the bridge rectifier which achieves full wave rectification to around 14 volts DC.

A capacitor may be used if the DC needs to be smoother. The DC is then put through a variable resistance in series with the battery to obtain the desired charging current. The charge time can be over several hours and always runs the risk of overcharging the battery.

Constant Voltage Charger:

Below I have drawn a very simple diagram of a **'Constant Voltage'** Battery Charger just to give you an idea of what is actually happening with the different charges.

Very similar to the Constant Current Charge, the voltage comes into the charger via a transformer which is 240 volt AC primary to around 15 volt AC secondary allowing for a slight voltage drop across the bridge rectifier which achieves full wave rectification to around 14 volts DC.

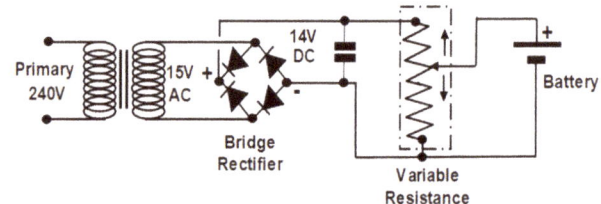

A capacitor may be used if the DC needs to be smoother. The DC positive is then put through a variable resistance in parallel to the battery to obtain the desired charging voltage. Full current is fed into the battery until the desired level is reached.

Sum up:

The **Constant Voltage** or **Constant Current** chargers with a simplified diagram above would be the one you would buy to keep in the garage for your car leaning more towards the **Constant Voltage** charger. A charger with a mixture of Constant Current and Voltage can be obtained cutting down the chances of overcharging.

Smart chargers may be used in industry for UPS systems where the battery banks have permanent contact with the charger and must not be overcharged. This is where float and trickle charge, explained later, comes into the equation.

Battery Documentation:

Description & Examples of: Permit to Work – Lesson Plan – Risk Assessment – Maintenance Form

General:

The Battery Maintenance Team and the documentation that they complete must be able to brief their Electrical Engineer in the condition of the batteries whilst they are located in an office quite a distance away.

The Electrical Engineer will be the person who determines what information that they require to ensure that the Maintenance System(s) that is in place is working and the batteries are being kept in good order. Documentation will differ from company to company in the information that is required.

At the same time there must be systems and documentation in place to ensure that personnel and plant are safe whilst the maintenance is carried out and that procedures that are in place are followed to the letter as with batteries there is no isolation.

I have given examples of a Risk Assessment, Lesson Plan and Maintenance Form, but I appreciate that these forms are just examples and some Electrical Engineers may require more or less information or documentation.

Permit to Work:

Don't think I have to outline much here as every company will have its own design of Permit. All I need say here is that Permits to Work are a way that the Plant Management keep control and knowledge of the work that is being carried out on their plant at any one time.

Lesson Plan:

This is sometimes called a Safe System of Work and it lists exactly how the maintenance is to be carried out safely. This can be a generic document that is issued as part of the documentation at the start of the work. It reminds the Maintenance Team of safety points and what to do in a certain order.

Risk Assessment:

A Risk Assessment is exactly what it says, so an estimate must be taken of what the dangers are and whether they can be removed. Some of these dangers may not be able to be removed such as batteries having to be worked on live as there is no way to isolate them. What precautions have been taken to ensure those remaining dangers are not a threat to life or plant? This must be completed by the Maintenance Team Leader before the work actually starts.

Maintenance Form:

This will differ enormously from company to company and is a record of the work that has been carried out, by whom and on what date. Some Electrical Engineers may require average readings where others may require more detailed results.

Battery Room Risk Assessment/Audit:

To be completed BEFORE any maintenance!

Battery Room Location: Date:

Name (Print):   Signature:

---

1 – Is there a Permit to Work issued?   YES/NO

2 – Has a Standard Lesson Plan been issued?   YES/NO

3 – Are all personnel on this maintenance competent?   YES/NO

4 – Is the room tidy and nothing untoward being stored within?   YES/NO

5 – Are all battery racks clearly numbered?   YES/NO

6 – Are all fire extinguishers clearly signed?   YES/NO

7 – Are all fire extinguisher examination labels in date?   YES/NO

8 – Are all eye wash stations clearly signed?   YES/NO

9 – Are all eye wash bottle solutions in date?   YES/NO

10 – Is there material present to deal with spillage?   YES/NO

11 – Are all items of PPE clean and stored correctly?   YES/NO

12 – Are all goggles of the secondary type and NOT primary?   YES/NO

13 – Are there notices up of how to deal with electric shock?   YES/NO

14 – Are there notices up on No Smoking or means of ignition?   YES/NO

15 – Are there warning signs of no food or drink?   YES/NO

16 – Are all liquids in containers clearly marked?   YES/NO

17 – Are all battery tools present and insulated?   YES/NO

18 – Are safety torches Atex Certified Exe?   YES/NO

19 – Are escape exits clearly signed with no obstructions?   YES/NO

20 – There should be at least **TWO** people together at any time.   YES/NO

NOTE: If one person leaves the room, work must stop!

21 – Are rubber mats on the floor?   YES/NO

22 – Are rubber sheets available to cover batteries in the vicinity?   YES/NO

23 – Please enter any comments in the box below:

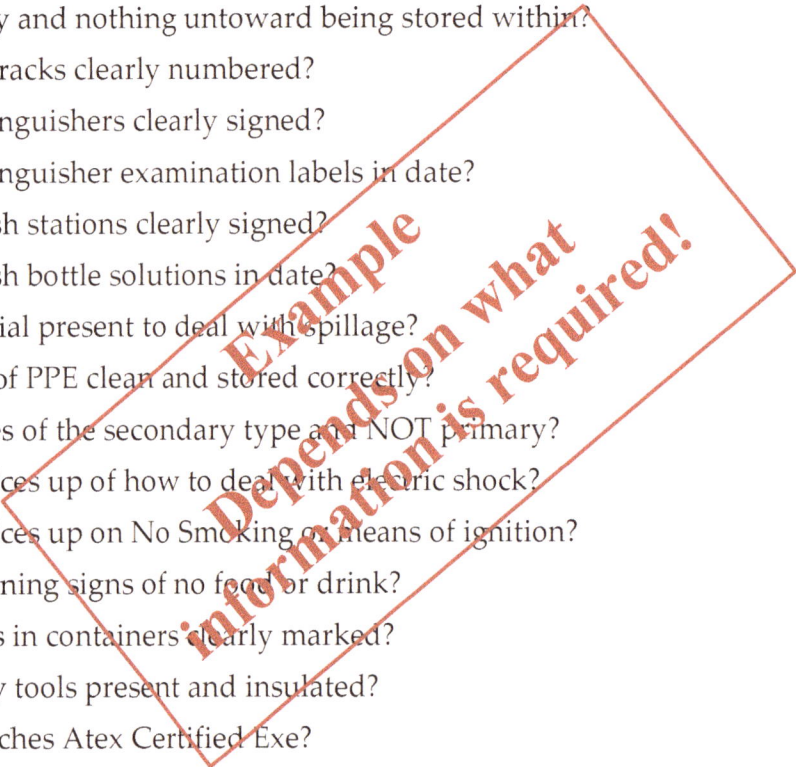

> Comments:

Battery Maintenance:

Form BM1

Form Number: Date:

Battery Room Location: Rack Number/Name:

How many Batteries in the Rack: Battery Type:

Electrolyte (i.e.: Acid/Alkaline):

Name (Print):   Signature:

Documentation:

1 – Has a Permit to Work been issued?   YES/NO

2 – Permit to Work Number: [   ]

3 – Has a Risk Assessment been completed?   YES/NO

4 – Has a Standard Lesson Plan been issued?   YES/NO

Isolations:

1 – With the charger still on are there any alarms up?   YES/NO

2 – If YES which alarms are up? [   ]

3 – Does the Thermal Image Device show any heat? (Charger ON)   YES/NO

4 – If YES where is the heat source? [   ]

5 – Switch off the charger to the Battery Bank as per works procedure. Check voltage reading.
                                                                         YES/NO

Batteries Heat:

1 – Are there any visible signs of heat?   YES/NO

2 – Does the Thermal Image Device show any heat? (Charger OFF)   YES/NO

3 – If YES where is the heat source? [   ]

Battery Corrosion:

1 – Are there any signs of corrosion on battery tops?   YES/NO

2 – If YES on what percentage of batteries in the bank? [   ]

Batteries Electrolyte Level:

1 – Having removed the caps is the electrolyte levels deemed to be low? YES/NO

2 – If YES what percentage of batteries in the bank? [   ]

3 – Top up the batteries so liquid level covers the plates to desired level.   YES/NO

4 – How much water in total was used to complete all batteries? [   ]

7 – Replace all caps and clean the tops of the batteries.   YES/NO

8 – Check tightness of all battery connections. (Report any that were slack.)   YES/NO

De-Isolation:

1 – Switch the battery charger back on. Check voltage reading.   YES/NO

Note: Specific Gravity to be taken of a percentage of batteries after a time period not straight after topping up with water!

105

Battery Maintenance Generic Lesson Plan: (Safe System of Work)
To be studied BEFORE any maintenance!

Battery Room Location: Date:

Name (Print):   Signature:

___

1 – Ensure that there is a Permit to Work issued.

2 – Ensure that a Lesson Plan has been completed.

3 – Ensure all maintenance personnel are in correct PPE.

    a – Goggles

    b – Apron

    c – Gloves

    d – Safety footwear

    e – Mask if applicable.

4 – Are ALL tools insulated?

5 – Ensure that all top up is filled with distilled/deionised water.

6 – Ensure that all Jewellery is removed or insulated. (Rings etc.)

7 – Ensure that the battery charger is isolated.

8 – Ensure the location of eye wash bottles.

9 – Ensure the location of fire extinguishers.

10 – Ensure the location of emergency exits.

11 – Ensure that signs are put at entrances 'Maintenance in Progress No Entrance!'

12 – Ensure the location of electrolyte spillage material.

13 – Ensure no personnel have food or drink in the battery room.

14 – All main personnel should be familiar with type of battery.

15 – There MUST be TWO maintenance personnel together at all times.

We cannot isolate batteries so systems are being worked on live.

16 – All personnel being trained should be supervised at all times.

17 – Ensure there are no means of ignition BEFORE removing caps.

18 – Ensure all safety torches are Atex Certified to Exe Standard.

19 – Some companies insist on rubber mats to stand on, others insist on rubber sheets to cover any batteries in the vicinity not being worked on.

INDEX

www.ingramcontent.com/pod-product-compliance
Lightning Source LLC
Chambersburg PA
CBHW041620220326
41597CB00035BA/6186